ADVANCED FOOD SCIENCE

Advanced Food Science

ROY HOPWOOD

Lecturer in Food Technology,
Blackpool College of Technology and Art

LONDON
G. BELL & SONS LTD
1975

India
Orient Longman Ltd
Calcutta, Bombay, Madras and New Delhi

Canada
Clarke, Irwin & Co. Ltd, Toronto

Australia
Edward Arnold (Australia) Pty Ltd, Port Melbourne, Vic.

New Zealand
Book Reps (New Zealand) Ltd, 46 Lake Road, Northcote, Auckland

East Africa
J.E. Budds, P.O. Box 4536, Nairobi

West Africa
Thos. Nelson (Nigeria) Ltd, P.O. Box 336, Apapa, Lagos

South and Central Africa
Book Promotions (Pty), Ltd, 311 Sanlam Centre,
Main Road, Wynberg, Cape Province

ISBN 0 7135 1908 8

Filmset by Typesetting Services Ltd, Glasgow, Scotland
Printed in Great Britain by J. W. Arrowsmith Ltd, Bristol

Preface

This book may be regarded as a natural sequel to the introductory work, *Elementary Food Science.* It attempts to consolidate the basic general scientific principles obtained by students who have either completed the preliminary work needed, or who, by virtue of 'O' levels in science, can appreciate food applications at a higher level.

The title, *Advanced Food Science,* is a relative one only; no attempt is made at specialisation in any food branch. Each chapter has a short number of questions with the answers to any numerical problems given at the end of the book. A list of suitable books for reference purposes is included. Background reading at this stage is highly recommended. Scientific terms have been defined where this has been thought necessary, any omissions can be rectified by reference to suitable paperback editions of food science dictionaries or science in general.

SI units are stressed but occasionally reference is made to other systems as the former are not fully operative at the present time in many of the smaller industrial concerns.

It would be impossible to cover the whole field of food in a book of this size, nor indeed has this been attempted; the items selected are those which are considered to be necessary for current examinations.

The general reader will be able to select those sections which are relevant to his or her future role in the food industry and to the syllabus requirements of the particular course being followed. It is believed that the book will be valuable for the following examinations in particular:

National Diploma in Baking;

Ordinary National Diploma in Food Technology; and

Advanced Certificate in Meat Technology.

Those students who are in the final year of a course of study leading to the award of *either* a Diploma in Meat Technology *or* an Advanced Bakery Craftsmen and Technicians Certificate will be more interested in the applications rather than a full knowledge of the science involved. Nevertheless, as a majority of these students continue to higher qualifications the theoretical aspects will assume greater importance.

It has been a difficult problem to decide what information will be of value for the Food Technicians Certificate (part 2). As a result of numerous visits to food firms involving discussions over their labour needs, the technical problems which arise, and the work needed on the marketable value of a product, I have found that a Food Technician would be expected to have a sound knowledge of food spoilage and food preservation. These items have received some detailed attention and a forecast made as to how industry views the future role of food preservation in this country and elsewhere. The majority of Food Technicians appear to be deeply concerned with the microbiology of food and less with the chemical and physical principles involved in its production. If this picture is correct a cursory glance at the latter will suffice and greater effort will be devoted to microbiological problems. In the last analysis it must be left to each individual to make his or her own choice.

In the collection of material I am again indebted to many friends and colleagues for their advice, and, in fairness to all, I have refrained from individual mention. I have also been grateful for the assistance of some strangers in other science departments and of several industrial firms. One regret is that sufficient space is only available to include a small fraction of the information they have liberally supplied.

I would like to thank the Institute of Meat, National Council for Bakery Education and National Council for Food Technology for permission to include a selection of papers covering the respective scientific principles involved.

R.H.

Blackpool,
March 1975

Contents

soaps and synthetic detergents—commercial cleaning
agents—in-line cleaning methods and flow requirements

1 Physico-chemical activity

PERIODIC TABLE

Initially the ninety-two natural elements from hydrogen to uranium were regarded as separate and unrelated substances. Early nineteenth-century chemists noted that sodium and potassium had similar chemical properties as did the triad of calcium, strontium and barium. An examination of the atomic weights of the latter elements shows a regular difference:

Calcium	40	
Strontium	87	47
Barium	137	50

Further investigation reveals similar atomic weight differences between the halogens (chlorine, bromine and iodine). As atomic weight determinations became more accurate an octave relationship appeared. If the elements were arranged in ascending order of atomic weight every eighth element resembled the element eight places before it, with minor reservations.

The final periodic table in its modern form may be explained in terms of the electronic configuration of each individual element. Elements which are chemically related have the same number of *external* valency electrons. Each element can only possess a certain maximum of electrons in each orbit or shell. The first shell may tolerate two electrons, the second shell eight, the third shell eighteen, etc. By reference to the first twenty elements (excluding hydrogen) the following pattern is revealed.

	He	Li	Be	B	C	N	O	F
Electrons	2	2:1	2:2	2:3	2:4	2:5	2:6	2:7
	Ne	Na	Mg	Al	Si	P	S	Cl
Electrons	2:8	2:8:1	2:8:2	2:8:3	2:8:4	2:8:5	2:8:6	2:8:7
	Ar	K	Ca					
Electrons	2:8:8	2:8:8:1	2:8:8:2					

Apart from the rare gases other vertical groups have the same number of valency electrons in the outermost shell. The horizontal columns are called series or, more commonly, periods. Members of each group of elements have related chemical and physical properties. Elements become increasingly basic as one descends a group and increasingly acidic as progress is made from left to right across a period. Sodium oxide is alkaline in nature which contrasts sharply with the acidic nature of sulphur dioxide. Potassium is more reactive with water than sodium, magnesium is more stable to water than calcium.

The periodic table in its complete form is useful in predicting the reactions of compounds used in the food industry, particularly where economic considerations are involved. Wherever possible sodium compounds are preferred to potassium as the former are considerably cheaper.

RADIOACTIVITY

Becquerel in 1896 showed that certain elements gave out radiation which could affect a photographic plate. He is principally remembered for his work on uranium. Mme Curie obtained radium from the mineral pitchblende and found this to be more active than uranium. Subsequent work has shown that elements of a greater atomic weight than bismuth (209) are all radioactive. To the spontaneous emission of radiation from these elements and their compounds, the term radioactivity was given.

Three distinct types of ray are involved:

1. Alpha rays—these consist of the nuclei of helium atoms.
2. Beta rays—These are high-velocity electrons.
3. Gamma rays—these are not material particles but are electromagnetic vibrations, characteristic of light, with a very short wavelength. They possess considerable powers of penetration; for example they can pass through 6 in. of steel. Like X-rays they are harmful to living tissue, both animal and vegetable.

Whenever an alpha particle is emitted a new element is produced which has an atomic number two less than the element from which it was formed:

$$\frac{226}{88} = \frac{222}{86} + \frac{4}{2}.$$

$$\text{radium} \qquad \text{radon} \qquad \text{alpha particle}$$

If a beta particle is lost then the atomic weight is unaltered. However, the atomic number increases by one to produce an element one place to the right in the periodic table:

$$\frac{234}{90} = \frac{234}{91} + \frac{0}{-1}.$$

$$\text{thorium} \qquad \text{proactinium} \qquad \text{beta particle}$$

The denominator to each symbol is the atomic number and the numerator is its mass number.

An element may have more than one different mass number or atomic weight on account of having different numbers of neutrons on the nucleus. Because they have the same arrangement of extra-nuclear electrons their chemical properties are identical and each atomic form is known as an isotope. Chlorine has three main isotopes of atomic weights 34, 35 and 37, which explains an average weight of 35.5 on a proportion basis.

A hazard of modern life is the 'fall-out' associated with nuclear explosions. Atoms of heavy metals of the uranium type are split into smaller particles or fission products. Over a period of time these unstable isotopes change into more stable ones. Fortunately only two isotopes exist long enough to be a source of worry to man's metabolism. Strontium 90 (isotope of strontium of mass 90) may be assimilated by grazing animals as a surface contaminant of grass. The body treats it in the same way as calcium, because of its periodic relationship, and deposits it in the bones. Its radioactivity can produce bone tumours or leukaemia. Caesium 137, which is in the same periodic group as sodium, can concentrate in the soft tissues of the body and affect genes.

Radioactive elements decay at different rates. In a certain period of time, one half of a portion of a radioactive element will disintegrate; in the next equal period, one half of the remainder (a quarter of the original amount) decays and so on. The period of time is known as the half-life period. For uranium, the half-life period is 4×10^9 years; for radium it is only 1.7×10^3 years and for some other elements it is a matter of seconds only.

Radiation applied under controlled conditions may be of advantage to the food industry. Goitre results if the iodine level of the diet falls to a minimum value and should be eliminated if the thyroid gland can absorb iodides and produce the hormone thyroxine. By administering radio-active iodine, and using a Geiger counter to follow the progress of the element in the body, a rapid diagnosis can be made of the efficiency of the thyroid gland.

Carbon 14 may be used by plants for photosynthesis purposes to produce nutrients that could result in malformed children as opposed to the normal effect of carbon 12.

Ionising radiations are thought to be of value in sterilising any food fresh or canned. Potatoes cease to sprout and the ripening of fruit is arrested. Tests have indicated that the milling and baking qualities of grain are unimpaired when doses of radiation large enough to kill insects are used. Micro-organisms are killed by the radiation energy from cobalt 60 and there is no evidence to suggest that the food will become radioactive to a dangerous extent. Insects, surface moulds and yeasts on oranges and destruction of Salmonellae in meat represent important victories for radiation. However, economically, other methods of control are far more competitive, and at the present time irradiation of food intended for human consumption is not legalised in Great Britain.

Electromagnetic radiation as a non-ionising radiation source is allowed in certain cases. Infra-red or heat rays can process food rapidly. Microwave ovens for heating pre-cooked foods prove very popular for quick snacks etc.

Ultra-violet rays attract flies to a lethal voltage source and are found in most plant bakeries. The shelf life of food may be extended with ultra-violet usage but is not encouraged because the cancer-producing effect of these rays on living animals could conceivably cause exposed food to become carcinogenic.

IONISATION

Atoms are regarded as electrically neutral particles because they contain equal numbers of electrons and protons. Ions are produced when electrons are lost or gained by a neutral atom or groups of atoms. Metals produce positive ions and non-metals negative ions.

$$Mg - 2e = Mg^{2+}$$
$$Cl + e = Cl^-$$

Inorganic acids, bases and salts form *strong* electrolytes in many cases and may have complete ionisation. In these cases only ions are produced. There are no undissociated molecules.

$$HCl = H^+ + Cl^-$$
$$NaOH = Na^+ + OH^-$$
$$NaCl = Na^+ + Cl^-$$

Most of the organic acids are *weak* electrolytes, citric, tartaric, succinic and propionic being examples of this type.

$$CH_3COOH \rightleftharpoons H^+ + CH_3COO^-$$

In the above equation for the dissociation of acetic acid there are only three ionised molecules out of every 1000 molecules present. Carbonic acid and ammonia solution are also poorly ionised.

$$NH_4OH \rightleftharpoons NH_4^+ + OH^-$$
$$H_2CO_3 \rightleftharpoons H^+ + HCO_3^-$$

Flavour is associated with ionisation, the degree of sourness or sharpness in acid based products depending upon hydrogen ion concentration. Yoghourt and carbonated beverages have characteristic flavours. Jelly strength increases with increasing hydrogen ion concentration until an optimum is obtained at about pH 3–3.3. Nearly neutral fruit juices will not produce a jelly with sugar because the sugar is not soluble enough at this pH to precipitate the pectin needed for stability. Thus the jelly stability depends upon a correct relationship between the acid, pectin and sugar concentrations.

Negative ions in equal concentrations can alter the temperature at which gelatine solutions for the meat industry set. Citrate ion will elevate the setting temperature whilst the iodide ion may lower the gelatin temperature below 0°C. The ability to control the setting temperature with flavour or nutritional advantages as a bonus results in an increased number of meat dishes.

Ionising radiations should not be confused with normal ionisation. The unit of radiation dose is the rad and represents an energy absorption of 100 erg/g of the substance irradiated.

Spoilage loss in strawberries can be reduced by using up to 300 krad (1000 rads = 1 kilorad or krad). Above this level there is a danger of the fruit softening and of reduced resistance to micro-organisms.

Gamma radiation on starch grains can alter their swelling facility and their solubility at different temperatures.

Ionisation occurs at all temperatures and is not confined to certain elements or compounds. It has not associated with it any of the potential health hazards of radioactivity.

OXIDATION

The term may be used to cover the following processes:

(*a*) addition of oxygen;

(*b*) loss of hydrogen;

(*c*) increase of valency;

(*d*) addition of any non-metallic radical or any non-metal (other than hydrogen); and

(*e*) loss of one or more electrons.

Reduction will be the *reverse* of any of the above processes. The processes of oxidation and reduction are complimentary and as a rule, one is always accompanied by the other. An oxidising agent is one which promotes oxidation and as a result is itself reduced in the reaction. Some of the common oxidising agents, with examples, are given below.

1. Oxygen

Combustion of the element in air

$$C + O_2 = CO_2.$$

2. Dioxides and peroxides

$$MnO_2 + 4HCl = MnCl_2 + Cl_2 + 2H_2O.$$

Manganese dioxide oxidises hydrochloric acid to chlorine and is itself reduced to manganese chloride

$$2KI + H_2O_2 + HCl = I_2 + 2KCl + 2H_2O.$$

Hydrogen peroxide may be used to estimate the amount of iodide ion in food as the free iodine produced can be volumetrically analysed with sodium thiosulphate solution.

3. Chlorine

In aqueous solution chlorine produces unstable hypochlorous acid which will liberate oxygen producing a bleaching action

$$H_2O + Cl_2 = HCl + HClO.$$

4. Potassium permanganate and potassium dichromate

Both reagents need acid conditions. Potassium permanganate is more widely used as the end point of the reaction, is clearer and in warm conditions rapid results are possible.

The oxidising agent can be represented generally by the following equation:

$$MnO_4^- + 8H^+ + 5e^- = Mn^{2+} + 4H_2O.$$

Note the electron transfer involved.

Strawberries and rhubarb can have their oxalic acid and oxalate content determined under laboratory conditions.

Potassium dichromate produces as one of its products a deep green chromium salt which masks the colour of any excess dichromate. Diphenylamine is used as an internal indicator and, when oxidised by a slight excess of potassium dichromate, produces an intensely coloured blue compound.

Ionically the oxidising power of potassium dichromate is shown as:

$$Cr_2O_7^- + 14H^+ + 6e^- = 2Cr^{3+} + 7H_2O.$$

Potassium dichromate is useful for the estimation of iron in food where chlorides may be involved as, unlike potassium permanganate, it is not affected by traces of chloride ion.

REDUCTION

As stated previously, reduction is the opposite of oxidation. A few of the reducing substances encountered in food technology are as follows.

1. Hydrogen

Certain metals will produce, in acid conditions, hydrogen which can effect reductions which do not happen if hydrogen is merely bubbled through the solution. However, ordinary hydrogen can reduce potassium dichromate to a chromium salt if cupric ions are present to act as a catalyst.

The production of lard compounds used as commercial shortening is by the process of hydrogenation. Hydrogen gas is passed into the heated oil at 180°C under an internal pressure of about 5 atm, with nickel as a catalyst. The unsaturated fatty acids are saturated by the addition of hydrogen at the double bonds.

$$-CH{=}CH- + H_2 = -CH_2-CH_2-$$

Fully hydrogenated whale oil which originally melts below 0°C can have a final melting point of 55°C to produce a stable solid at room temperature.

The linoleic acid content of the triglycerides used in biscuit fat should be kept as low as possible by selective hydrogenation. A shelf life of several years can be achieved using air-proof tins and allowing the natural anti-oxidants in flour to exert their effect.

'Hydrogen swell' causes canned fruit to alter in flavour and texture as well as producing blown cans.

2. Metals

Most metals have a potential reducing value in foods. Traces of copper in particular can alter food colour and the use of sequestering agents to 'lock-up' the offending metal into a dormant condition is needed. The higher a metal is in the electrochemical series the greater is its reducing effect.

3. Sulphur dioxide

Fruit pulp is preserved using a concentration of 1000–1500 p.p.m. The main advantage of using sulphur dioxide is that with water it forms sulphurous acid.

$$H_2O + SO_2 = H_2SO_3$$

In the boiling process most of the gas is removed from the unstable acid. Legal standards are laid down for the amount of residual gas in jam. Strawberry jam in particular needs periodic checking as the fruit is very retentive of the preservative.

4. Sodium thiosulphate

This compound is used extensively in volumetric analysis to estimate iodine. Wijs solution based upon iodine monochloride gives information about the chemical nature and freshness of edible oil and fat samples when any free iodine is titrated against standard sodium thiosulphate solution.

$$2S_2O_3^{2-} = S_4O_6^{2-} + 2e^-$$
tetrathionate ion

REDOX POTENTIAL

The close relationship between oxidation and reduction has already been indicated. The term redox reaction refers to a chemical reaction in which an oxidising agent is reduced and a reducing agent is oxidised. Redox is therefore an abbreviation of reduction–oxidation. All such reactions can be related to the energy needed to cause electrons to be transferred from one atom, ion or molecule to another.

Hydrogen is taken as a reference point.

$$\tfrac{1}{2}H_2 = e^- + H^+$$

The energy needed to produce a hydrogen ion by the loss of one electron is arbitrarily taken as zero volts. All other elements, either as neutral atoms or in different valency states, can have their ionisation potential referred to hydrogen.

Figure 1 shows a few of these changes. An element or ion charged negatively with respect to hydrogen has a greater tendency to lose electrons, the reverse being true for a positive reading. The respective merits of free atoms and ions as reducing or oxidising agents can also be ascertained if one regards oxidation and reduction as electron transfer processes.

	Reaction system	Volts
	$K = e^- + K^+$	-2.92
	$Na = e^- + Na^+$	-2.71
Strength as reducing agent decreases	$Fe = 2e^- + Fe^{2+}$	-0.44
	$\frac{1}{2}H_2 = e^- + H^+$	**0.00**
	$Cu = 2e^- + Cu^{2+}$	$+0.34$
Strength as oxidising agent increases	$Fe^{2+} = e^- + Fe^{3+}$	$+9.76$
	$Cl^- = e^- + \frac{1}{2}Cl_2$	$+1.36$

Fig. 1.   Redox series

Dough is a complex redox system. As fermentation proceeds compounds are oxidised and reduced. Potassium bromate and ascorbic acid, if added as improvers, alter the *status quo* of the system.

A simpler example may be found in the colour changes associated with meat, and the actual compounds involved will be discussed later. The ferrous ion occurs in fresh meat and the ferric ion in old meat. These represent two different states of oxidation of iron.

$$Fe^{2+} \quad = e^- + \quad Fe^{3+}$$

reductant or   oxidant or
reduced form   oxidised form
(fresh)   (old)

ELECTROCHEMICAL SERIES

Electricity in the simple cell is a result of electrons flowing externally from a copper to a zinc electrode when both electrodes are immersed in a suitable electrolyte. Metals immersed in water or an electrolyte produce positive ions and leave an excess of electrons on the metal surface. Because of the attraction of opposite charges, the ions remain close to the metal and an electric double layer is produced. At equilibrium the rate at which ions are produced equals the rate of ion deposition on the metal. Thus, for a divalent metal like zinc:

$$Zn \rightleftharpoons Zn^{2+} + 2e^-.$$

The double layer results in a potential difference or *electrode potential* between the surface of the metal and the liquid. All metals can have their *relative* electrode potentials measured if hydrogen forms the second electrode in a circuit. Hydrogen is allowed to bubble over a platinum surface and is adsorbed. The result is that the electrode behaves as a solid hydrogen electrode and the following equilibrium is established:

$$H_2 - 2e^- \rightleftharpoons 2H^+.$$

As stated previously, under redox conditions the electrode potential of the hydrogen electrode is taken as zero. Under standard electrolyte conditions, using a potentiometer, the electrode potentials of the elements can be obtained. An abbreviated table is shown in Fig. 2.

↑ Reducing agents	$Na(Na^+)$	-2.71	Solution pressure
	$Zn(Zn^{2+})$	-0.76	$>$ ionic pressure
	$Fe(Fe^{2+})$	-0.44	$\therefore$ metal surface has excess electrons
↓ Oxidising agents	$\frac{1}{2} H_2(H^+)$	**0.00**	**NEUTRAL**
	$Cu(Cu^{++})$	$+0.34$	Ionic pressure
	$Ag(Ag^+)$	$+0.80$	$>$ solution pressure
	$Au(Au^+)$	$+1.68$	$\therefore$ ions deposit on metal surface giving a positive charge

Fig. 2. Electrochemical series (electrode potential in volts).

The resulting potential difference of any combination of metals can be obtained by subtracting algebraically their standard electrode potentials:

$$Cu/Zn \text{ electrodes} = 0.34 - (-0.76) = 1.1 \text{ V.}$$
$$Ag/Cu \text{ electrodes} = 0.80 - 0.34 \quad = 0.46 \text{ V.}$$

In all cases the standard electrode potentials of elements and ions are also their standard redox potentials. When the periodic table and the electrochemical series of the elements are closely examined many important conclusions can be reached.

1. The nature of electrolytic products may be predicted. Brine solution contains four ions, only two of which are discharged at the appropriate electrode.

$$NaCl \rightleftharpoons Na^+ + \underline{Cl^-}$$
$$H_2O \rightleftharpoons \underline{H^+} + OH^-$$

Hydrogen ions are discharged at the cathode in preference to sodium ions as they have a lower discharge potential. With gases the discharge potential of the corresponding ions may differ considerably from their electrode potential. To discharge an OH^- ion and produce oxygen gas requires a larger amount of energy than is obtained from its standard electrode potential. The difference between the two is called overvoltage or overpotential. Due to overvoltage the OH^- ion has a higher discharge potential than the Cl^- ion and in consequence the latter is discharged and chlorine gas is produced.

2. Combination of metals with oxygen: sodium oxide is more stable to heat than mercuric oxide, mercury being produced at a fairly low temperature. The affinity of metals for oxygen decreases down the series. It is impossible to reduce the oxides of the alkali metals with carbon, whereas zinc and lead can be produced from their oxides by this technique.

Corrosion is akin to oxidation and the canning industry requires metal surfaces to be non-toxic and stable under acid conditions. Zinc occurs above tin in the series and is used to coat iron or steel (galvanisation).

Galvanised iron when scratched so as to expose the iron will not rust as zinc ions are produced rather than ferrous ions. Stainless steel and galvanised containers may be used to hold food with a low acidity or near neutral pH. Tinned iron is preferred for holding fruit since tin is less reactive than zinc or iron in acid conditions due to its low position in the series. However, if tinned iron is scratched and the iron exposed, rusting is accelerated as ferrous ions are produced rather than stannous ions. Lacquering of cans has the added virtues of limiting damage by scratching or colour changes due to pH alteration if metal ions are produced.

3. Displacement of elements: normally a metal is displaced from aqueous solutions of its salts by a metal which is higher in the electrochemical series. The further apart the metals are, the more rapid is the displacement action. Iron will displace copper from copper sulphate solution and zinc will displace both metals from solutions of their salts.

$$Fe + Cu^{2+} = Fe^{2+} + Cu$$
$$Zn + Fe^{2+} = Zn^2 + Fe$$

Aluminium behaves anomalously due to a thin, but resistant, film of surface oxide and in consequence has poor displacement powers.

Halogens are displaced from their solutions by other halogens lower in the series. Bromine and iodine, although small in quantity, are valuable elements in food or food additives and can be concentrated from dilute solutions using chlorine.

$$Cl_2 + 2Br^- = 2Cl^- + Br_2$$
$$Cl_2 + 2I^- = 2Cl^- + I_2$$

COLLOIDS

Our previous knowledge has shown that the properties of colloids depend upon several features, one of which is particle size. The difference in size between a particle in suspension and one in colloidal solution means an enormously greater surface area per unit mass of the latter. Figure 3 shows some of the characteristic differences of true solutions, sols and suspensions.

The two types of sols are lyophobic (solvent hating) and lyophilic (solvent loving). When the dispersion medium is water, the sols may be described as hydrophobic and hydrophilic. Lyophobic sols have little

True solutions	*Gels*	*Suspensions*
Molecular subdivision	Colloidal subdivision	Mechanical subdivision
Particles are not visible with ultra-microscope	Refracted light of particles is visible with ultra-microscope	Particles visible in glass microscope or naked eye
No gels produced	Gels always formed	No gels produced
Transparent particles pass through semi-permeable membranes	Transparent particles retained by semi-permeable membranes	Generally opaque particles retained by filter paper
Systems show high osmotic pressure	Systems show low osmotic pressure	Systems show no measurable osmotic pressure
Intense kinetic activity	Brounian movement	Little movement

Fig. 3.

attraction between disperse phase and the medium, whilst the reverse is true of lyophilic sols.

In food colloidal systems the distinction between hydrophilic and hydrophobic sols is a fine one. A slight physical change may alter a given colloid from one group to the other. Egg albumen is a hydrophilic colloid, being extremely soluble in water. However, when coagulated by heat it becomes insoluble in water and therefore hydrophobic.

Protein denaturation invariably results in the promotion of a hydrophobic system. Hydrophilic colloids include agar-agar, starch, gelatin, pectin and native proteins. A protein or fatty acid molecule may contain a mixture of polar (electrovalent groups) and non-polar (covalent groups). The efficiency of emulsifiers and stabilisers in foods depends upon a balance between polar and non-polar groups.

Many of the methods and ingredients used in food preparation can produce increased or decreased dispersion of colloidal particles. Homogenisation of cream or milk is a mechanical means of increasing the dispersion of fat particles. However, although beating a curdled custard may reduce particle size, it cannot change completely a suspension system to one of colloidal proportions.

The production of acid during the fermentation of bread dough induces dispersion of the gluten, but in milk causes the casein to form a clot. If enough acid is added the casein clot may be dispersed.

The addition of acid to proteins alters the pH in relation to the isoelectric point of the protein (isoelectric point will be mentioned in the next chapter).

In cakes, the addition of soda in excess of the amount needed to neutralise the normal acidity of the mixture may increase gluten dispersion producing a definite grain or crumb texture.

Alkalies cause fruits and vegetables to lose their firm texture and become mushy. It is believed that this is due to the greater dispersion of the cellulose and pectic substances. The temperature of egg coagulation is elevated when alkalies are added. The addition of alkali to dough produces a greater gluten dispersion and the dough becomes runny and sticky to handle. The flavour may be affected and in larger quantities the baking quality of the flour is impaired. Changes of pH also affect the physical properties of proteins. At the isoelectric point the swelling of gelatin is at a minimum as are its viscosity, stability in solution, osmotic pressure and ash content. The setting of gelatin is delayed by increased pH and in dried *legumes* the effect is to disintegrate some of the protein.

Enzymes are capable of producing dispersion changes in food. Milk clotting is due to the addition of rennin, which contrasts strongly with the increased gluten dispersion produced by proteinase in flour.

Colloidal systems in food are numerous as the dispersed phase may be solid, liquid or gaseous and the dispersing medium liquid or solid. Ice-cream has a gaseous dispersed phase but a solid dispersing medium, whilst mayonnaise is essentially in the liquid condition throughout. However, in all cases the properties of any colloidal system are fundamentally related to the very large area of the dispersed phase.

EMULSIONS

Vegetable oils and water are, under normal conditions, incapable of mixing. Agitation and heat may produce a temporary alliance but immiscibility results when the original conditions are restored. It is possible, though, to disperse one liquid in small droplets into another and produce an apparently stable mixture which is known as an emulsion.

Two types of emulsion can be recognised according to whether the water or the oil is the dispersed phase, that is, water-in-oil (W/O) or oil-in-water (O/W). To achieve stability, a third substance, an emulsifier or emulsifying agent, is necessary. Milk is a natural oil-in-water emulsion, stabilised, as it is, by the protein casein, which as a thin film surrounds each fat globule and prevents coalescence.

The type of emulsion formed (that is, O/W or W/O) depends upon several factors, among which are:

(*a*) nature of the oil;
(*b*) proportion of oil and water;
(*c*) presence of electrolytes;
(*d*) order of incorporation of the oil and water; and
(*e*) nature of the emulsifying agent.

Homogenisation techniques are designed to produce a disperse phase of near colloidal dimensions and the emulsifying agent lowers the surface tension of both liquids and acts as a link between them. In general, hydrophilic emulsifiers favour the formation of O/W, whereas hydrophobic ones favour a W/O system.

Emulsion classification can be determined using a microscope. A drop of the emulsion is placed on a slide, and while it is being observed through a microscope a small drop of water is added and stirred with a pin point. If the water blends with the emulsion it is an O/W emulsion. However, if the oil blends with another drop, the outside phase is oil and it is a W/O emulsion. In general, an O/W emulsion conducts an electrical current readily, whereas a W/O does not.

Food emulsifiers must be edible, preferably odourless, light in colour and cheap. An emulsifier should have a good balance of both polar and non-polar groups. The polar parts are attracted to the water phase whereas the non-polar groups will be orientated to the oil phase. Emulsifying agents with a low hydrophilic/high lipophilic balance are ideal for stabilising W/O emulsions, whereas emulsifying agents with a high hydrophilic/low lipophilic value stabilise O/W emulsions.

A large number of emulsifying agents have been prepared, taking advantage of their properties to produce a number of different functions in foodstuffs. The following list illustrates a selection of those important in the baking industry.

1. Bread improvers: these may be used in the form of pastes, creams and powders. The result is better crumb structure, increased volume and retardation of staling. In specialised items such as French stick bread and rolls, they can produce large volume and a crisp, attractively coloured crust.

2. Cake improvers: emulsifying agents are essential for the preparation of high ratio cakes where a high level of moisture and sugar are used in relation to the flour. In addition, they contribute to good aeration, texture and volume.

3. Fillings: it may be necessary to retard the process of sugar crystallisation in a range of selected fillings. Examples include almond paste, white bean paste and syrup waffles.

A characteristic meat emulsion is sausage in which the protein and water act as a dispersing medium and fat as the dispersed phase. A meat emulsion has, therefore, characteristics related to an oil-in-water emulsion. The primary emulsifying agent in a meat emulsion is myosin from the muscle. The addition of salt to a sausage product assists in releasing myosin from the muscle fibre. Over-chopping of the fat should be avoided as the surface area of the fat is increased to such an extent that the water–protein phase cannot hold the fat in its highly dispersed condition.

The addition of phosphate assists in the incorporation of high levels of fat or moisture in sausage products. It is held by some authorities that polyphosphates are not true agents, but rather emulsion stabilisers, in that they have a buffering action and maintain the pH at an optimum level for emulsion stability. However, it is generally accepted that the ultimate result is greater retention of fat and moisture in a cooked product. Cereals containing lecithin, and milk products with casein, may also be used in meat products as extenders and emulsifiers.

Food emulsifiers may include egg yolk, whole egg, milk, pectin, gelatin, starch pastes and glyceryl mono-stearate (G.M.S.). Food stabilisers are illustrated by gums and alginates. The lists are extensive and many firms market products which combine both features. An example would be the ester produced from propylene glycol and alginic acid, that is, propylene glycol alginate. In food emulsions it is unaffected by polyvalent ions or low pH values and combines the other desirable features of such additives in being economical and safe in food products.

Emulsifiers have played an important part in the preparation of margarines, shortenings, ice-cream and toffees such as caramels. We may add to this list such items as bottled chocolate drinks, coffee whiteners, sauces, mayonnaise, peanut butter, convenience foods, icings, artificial creams and even chewing gum.

The food technologist is available to be consulted on the right choice of emulsifier for any particular food product.

EXERCISE 1

1. In what ways does radioactive disintegration differ from ordinary chemical reaction? Explain briefly what is meant by:
    (a) a radioactive series;
    (b) half-life period;
    (c) an isotope.
2. What is a food emulsion? Indicate any precautions necessary to achieve stability if quality of a bulk manufactured product is to be maintained.
3. How does the electrochemical series account for the reactivity of food elements? Illustrate your answer by the selection of specific elements used in the food industry either by themselves or in the form of compounds.
4. What do you understand by the terms:
    (a) oxidation and reduction;
    (b) colloidal systems? Illustrate your answer by reference to substances of interest to the food technologist.
5. What physical and chemical tests could be used to distinguish a colloidal from a molecular solution? How could one regard:
    (a) proteins;
    (b) mineral food additives;
    (c) carbohydrates as regards their solution properties.
6. What is an ion, and what classes of substances are ionised in aqueous solution? Indicate briefly any advantages or disadvantages which might result if ions are present in food mixtures.

2 Chemical and physical problems

FORMULAE TYPES AND WEIGHT PROBLEMS

Inorganic compounds have a molecular formula which expresses adequately the composition of that substance. The quantitative analysis of an organic compound would produce a molecular formula which might not reveal the exact chemical nature of the compound in question. When the percentage of each element is divided by the atomic weight of that element, and the ratio is then expressed in whole numbers by dividing each term by the lowest value, or by some simple fraction of this value, an *empirical* formula is obtained. This formula is merely the simplest expression of the ratio of the atoms in the molecule, as the following example shows:

A substance has the following composition $C = 39.9$, $H = 6.7$, $0 = 53.4$

$$C = \frac{39.9}{12} = 3.33$$

$$H = \frac{6.7}{1} = 6.7$$

$$O = \frac{53.4}{16} = 3.33$$

Dividing through by 3.33, and allowing for experimental error, the ratio of the atoms $C:H:O = 1:2:1$, therefore the empirical formula is CH_2O.

In order to determine the correct *molecular* formula which would yield the actual number of atoms in the molecule, the molecular weight of the compound must be determined. If, for example, the molecular weight was 60 then its molecular formula is $(CH_2O)_2$ and not CH_2O or $(CH_2O)_3$, etc.

By reference to the valency of each element or radical involved, and by proving the latter by selected chemical tests, the *structural* formula emerges. This is the most important formula and shows the type of bonding present.

If the compound contained only a methyl (CH_3) and a carboxylic $(COOH)$ group then its structural formula would be:

15

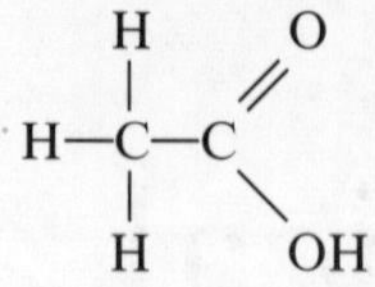

acetic acid

In routine analysis the final structural formula may be deduced from weight observations and physical properties.

Example

A very volatile liquid with a pleasant, fruity odour like apples is subjected to analysis. Qualitatively, the elements present included carbon, hydrogen, oxygen and nitrogen. 0.354 g yielded on combustion 0.415 g carbon dioxide and 0.212 g water. A Kjeldahl determination revealed 18.5 per cent nitrogen. Vapour density measurements indicated a molecular weight of 75. What structural formula can be assigned to the substance?

Solution

$$0.415 \text{ g of } CO_2 \text{ contain } \frac{3}{11} \text{ of } C = 0.113 \text{ g}$$

$$0.212 \text{ g of } H_2O \text{ contain } \frac{1}{9} \text{ of } H = 0.0235 \text{ g}$$

$$\therefore \text{ the percentage of C in the compound} = \frac{0.113 \times 100}{0.354} = 31.9 \text{ per cent}$$

$$\therefore \text{ the percentage of H in the compound} = \frac{0.0235 \times 100}{0.354} = 6.6 \text{ per cent}$$

∴ the percentage of oxygen in the compound by difference

$$= 100 - (31.9 + 6.6 + 18.6) = 42.9 \text{ per cent}$$

The empirical formula is $C_2H_5NO_2$, this is also the molecular formula in view of 75 obtained for the latter. Two possible structural formula exist:

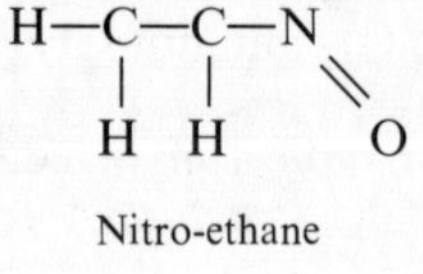

Nitro-ethane   or   Ethyl nitrite

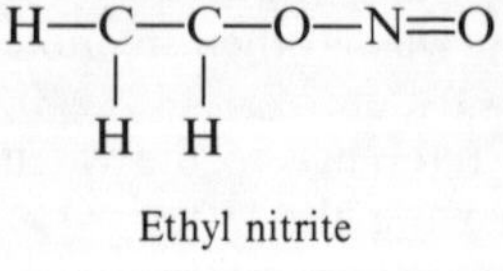

Nitro-ethane boils at 114°C and ethyl nitrite at 16°C. Odour and boiling point clearly indicate the compound to be ethyl nitrite and *not* nitro-ethane.

Ethyl nitrite when diluted with alcohol is used medicinally as 'sweet spirit of nitre' since it possesses stimulant properties. The odour resembles certain organic synthetic flavours used in food, known as *esters*.

By employing other physical and chemical tests, the identity of substances found or added to food can be readily established on a weight/formula basis.

ATOMICITY

The molecule has been referred to as the smallest unit of an element that can exist under normal conditions. Simple gases such as hydrogen, oxygen and nitrogen always have two atoms per molecule and are said to be *diatomic* molecules. By definition, the atomicity of an element is the number of atoms in one molecule of it.

The rare gases, for example, helium, contain only one atom in the molecule and are *monoatomic*. Separate atoms of the simple gases do not normally exist. Atomic hydrogen is an extremely active form of the element and has powerful reducing potential. Ozone has three atoms per molecule and is therefore *triatomic*. Other elements important in food, such as sulphur and phosphorus, contain more than three atoms per molecule and are termed *polyatomic*. The correct molecular formulae, from vapour density measurements, assigns to these elements S_8 and P_4 respectively.

As the molecular weight of a gas or vapour is twice its vapour density, a knowledge of atomicity enables the correct identification of chemical compounds to be easily determined.

VAPOUR DENSITY

The density of a gas may be expressed in two ways. First, it may be stated as normal density in g/cm^3 at a standard temperature and pressure. Secondly, its density relative to a standard gas, usually hydrogen, is often preferred. The vapour density of any gas may, therefore, be defined as the number of times a given volume of a gas is as heavy as an equal volume of hydrogen, both volumes being measured at the same temperature and pressure.

Determinations of vapour density are valuable for the calculation of molecular weights of elements or compounds which can be vaporised without decomposition. As with gases, it is found that:

Vapour density × 2 = approximate molecular weight of the substance in the vapour condition.

Figure 4 gives the exact values of the normal densities of gases which are associated with the food industry. The rate of diffusion of a gas is related to its vapour density, and is thus of importance in flour bleaching.

Hydrogen	0.09	Neon	0.90
Methane	0.72	Nitric oxide	1.34
Ammonia	0.77	Oxygen	1.43
Carbon monoxide	1.25	Carbon dioxide	1.98
Nitrogen	1.25	Sulphur dioxide	2.93
Air	1.29	Chlorine	3.21

Fig. 4.   Normal densities of gases.

As with density problems involving liquids and solids, it is much easier to refer to the relative density of a gas than its absolute density. The relative density of air to hydrogen is just over 14 and may be used in industrial problems involving gaseous products.

AVOGADRO'S HYPOTHESIS

By combining a knowledge of the density of a gas and its volume change on decomposition or reaction with other substances, the composition of some gaseous compounds may be determined by using Avogadro's hypothesis. He stated that 'equal volumes of all gases under the same conditions of temperature and pressure contain the same number of molecules'.

Like a theory, an hypothesis is an idea put forward to explain a fact. It has been consistently correct and is sometimes referred to as Avogadro's law.

The number of molecules in 1 g molecule of any substance, solid, liquid or gas, has been accurately determined experimentally by several independent methods. This number, called the Avogadro number, is the same in every case and is 6.023×10^{23}. Therefore, in 1 g molecule of any gas (that is in 22.4 litres at S.T.P.) there is this number of molecules.

Avogadro's hypothesis is applied to the solution of chemical problems involving gas volumes, particularly in the field of combustion.

Example

What volume of air would be required for the complete combustion of a mixture of 50 cm³ of methane, 20 cm³ of hydrogen and 40 cm³ of carbon monoxide? (Air contains 21 per cent by volume of oxygen.)

The equations are written down and underneath are the volume relationships as deduced from Avogadro's hypothesis.

$$CH_4 + 2O_2 = CO_2 + 2H_2O$$
1 vol   2 vol

$$2H_2 + O_2 = 2H_2O$$
2 vol   1 vol

$$2CO + O_2 = 2CO_2$$
2 vol   1 vol

Gas volumes are now replaced by actual quantities.

50 cm³ of methane combine with 100 cm³ of oxygen
20 cm³ of hydrogen combine with 10 cm³ of oxygen
40 cm³ of carbon monoxide combine with 20 cm³ of oxygen

The total volume of oxygen needed for the combustion of all three gases is $100 + 10 + 20 = 130$ cm³.

$$\text{Volume of air required} = \frac{130 \times 100}{21} \text{ cm}^3$$
$$= 619 \text{ cm}^3.$$

GAS VOLUME CALCULATIONS

Volumes of gases are usually measured under the temperature and pressure conditions of a particular experiment. For purposes of comparison these volumes must be reduced by use of the general gas equation to S.T.P.

The addition of chalk to flour for mineral additive purposes may be estimated quantitatively by noting the amount of gas released and hence the purity of the sample.

Example

5 g of impure chalk, when dissolved in excess of dilute hydrochloric acid, gave 588 cm³ of carbon dioxide at 780 mm mercury (or barometric) pressure and 21°C. What is the percentage weight of calcium carbonate in the chalk? ($Ca = 40$, $C = 12$, $O = 16$, gramme molecular weight of a gas occupies 22.4 litres at S.T.P.)

The equation for the reaction is:

$$CaCO_3 + 2HCl = CaCl_2 + H_2O + CO_2.$$
$$100 \qquad\qquad\qquad\qquad\qquad 44$$

The general gas equation now converts the volume to S.T.P.

$$\frac{780 \times 588}{294} \times \frac{760 \times V_1}{273}$$
$$\therefore V_1 = \frac{780 \times 588 \times 273}{294 \times 760} = 560.6 \text{ cm}^3.$$

The gramme molecular weight of carbon dioxide is 44.

$\therefore 22400$ cm³ of CO_2 at S.T.P. weighs 44 g
$\therefore 560.6$ cm³ of CO_2 at S.T.P. weighs $\dfrac{44 \times 560.6}{22\,400} = 1.102$ g.

From the equation:

100 g of pure chalk would yield 44 g carbon dioxide

$\therefore$ 5 g of pure chalk would yield 2.2 g carbon dioxide.
However the sample only yielded 1.102 g of carbon dioxide

$$\therefore \text{ percentage purity of sample} = \frac{1.102 \times 100}{2.2} = 50 \text{ per cent approx.}$$

In the meat industry the permitted level of sulphur dioxide is based upon a weight/volume relationship of sodium metabisulphite/sulphur dioxide conversion.

The addition of chlorine in drinking water is of the order of 1–3 p.p.m. and at a maximum of 11 p.p.m. for water in the canning industry. At these low levels of gas addition a volumetric rather than a gravimetric approach is preferred. By using appropriate formulae it is easy to calculate what weight of chlorine compound might be needed to give a desired result. Liquid chlorine is used in the meat industry for addition to water. This however involves stringent safety regulations which would not apply to bleaching powder.

ASSOCIATION AND DISSOCIATION

In certain cases, particularly where high temperatures are involved, the vapour density obtained gives a molecular weight different from the expected value. When less than expected, the abnormality is due to *dissociation*. In examples involving a higher value than expected the term *association* is used.

Aluminium chloride exists as Al_2Cl_6 molecules at 400°C. At a temperature of 1100°C it dissociates to give $AlCl_3$ molecules.

$$Al_2Cl_6 \rightleftharpoons 2AlCl_3$$

At higher temperatures still, it dissociates into chlorine and aluminium.

Association is often related to substances which contain hydroxyl groups, alcohols and water being typical examples. Water shows a very high boiling point for a compound with a low molecular weight of 18 (Ammonia of molecular weight 17 boils at -33°C). In addition, the specific heat and density variations of water near the freezing point are compatible with an increase in molecular complexity.

In benzene solution, both acetic and benzoic acids are associated into double molecules, but in water solution (apart from a very small dissociation of acetic acid) they are present in single molecules. The distribution of either acid in a preservative system involving a covalent/electrovalent system can be investigated by titration techniques involving standard alkali and using phenolphthalein as indicator.

Ammonium salts and nitrogen dioxide are good examples of dissociation. Figure 5 shows the effect of heating some ammonium chloride in a combustion tube.

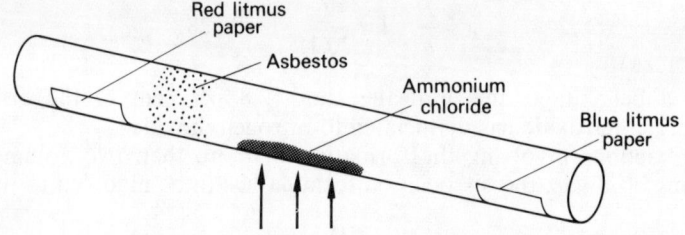

Fig. 5. Disassociation of ammonium chloride.

Due to its smaller molecular weight, the ammonia formed by dissociation diffuses through the asbestos plug more rapidly than the heavier hydrogen chloride.

$$NH_4Cl \rightleftharpoons NH_3 + HCl$$

As a result, the litmus paper above the plug turns blue (excess alkali) and the litmus paper below the plug turns red (excess acid).

Nitrogen dioxide can be liquefied using a freezing mixture of ice and salt at normal pressures. As the temperature rises to about 20°C the liquid boils and produces the familiar brown colour of nitrogen dioxide gas. The colour deepens up to 140°C and then gradually fades until at about 620°C no colour is visible. Two dissociations have occurred which can be accounted for as follows:

$$N_2O_4 \quad \rightleftharpoons \quad 2NO_2 \quad \rightleftharpoons 2NO + O_2.$$

very pale   deep brown   colourless
yellow liquid
(dinitrogen tetroxide)

The degree of dissociation in any system where there is an increase from one volume to two volumes on complete dissociation can be calculated from the following formula.

$$x = \frac{D}{d} - 1$$

where x = degree of dissociation, D = density for no dissociation and d = observed density.

Example

The vapour density of nitrogen dioxide at 60°C is found to be 30.1. What information does this give as to the state of the nitrogen dioxide molecules at this temperature?

Dinitrogen tetroxide is dissociating to nitrogen dioxide as the temperature rises, until at 140°C only nitrogen dioxide molecules are present. The vapour density of undissociated dinitrogen tetroxide is 46.

$$\therefore \; x = \frac{D}{d} - 1 = \frac{46}{30.1} - 1 = 0.528.$$

As a percentage, this indicates that 52.8 per cent of the original dinitrogen tetroxide has dissociated to nitrogen dioxide.

In reactions involving the formation of more than two volumes of gaseous products, the dissociation formula has to be modified to read:

$$x = \frac{D - d}{d(n-1)}$$

where n = the number of gaseous molecules produced from one original molecule.

The dissociation of ammonium carbamate in biscuit aeration yields two molecules of ammonia and one molecule of carbon dioxide.

$$NH_2COONH_4 \rightleftharpoons 2NH_3 + CO_2$$

OSTWALD'S DILUTION LAW

Inorganic acids, alkalis and salts will easily conduct electricity due to their complete dissociation into ions. A large number of organic compounds will not conduct electricity at all. They are incapable of ionisation. There are therefore two distinct groups, good conductors or electrolytes and non-conductors or non-electrolytes.

Organic acids in food have conducting properties which lie between the previous two extremes.

Except under conditions of high concentration, strong acids, strong bases and nearly all salts are wholly dissociated into ions. A large number of organic compounds including flavour additives, emulsion stabilisers and anti-oxidants produce few ions, and Ostwald presumed that such molecules would be relatively unaffected by dilution as regards their electrical conductivity.

The dissociation constant for a binary electrolyte depends upon the amount of dissociation reached at the equilibrium point, and for weak electrolytes, such as acetic acid, the value obtained should, theoretically, be independent of dilution, but will vary with temperature.

Ostwald obtained the following expression which can be tested experimentally over a wide dilution range.

$$K = \frac{M^2}{(1 - M)V}$$

where K = dissociation constant, M = degree of dissociation and V = volume of solution in litres. The expression is known as Ostwald's dilution law for a binary electrolyte.

Example 1

$\dfrac{N}{100}$ acetic acid has a degree of dissociation of 0.042 at 18°C. What is the equilibrium constant?

$$K = \frac{M^2}{(1-M)V} = \frac{(0.042)^2}{0.958 \times 100} = 1.84 \times 10^{-5}$$

Example 2

At 18°C decinormal acetic acid is 1.35 per cent ionised. Calculate the equilibrium (dissociation) constant.

$$K = \frac{M^2}{(1-M)V} = \frac{(0.0135)^2}{0.9865 \times 10} = 1.85 \times 10^{-5}$$

The values of K in both cases are practically identical, showing that acetic acid obeys the law.

If however, the equilibrium constant for 1 g mol of sodium chloride is calculated, under the same conditions of temperature and dilution factor, no constant is obtained. $\dfrac{N}{10}$ sodium chloride gives $K = 0.48$ and $\dfrac{N}{100}$ sodium chloride reveals $K = 0.147$.

Ostwald's dilution law may be used to find the pH of week acids since they will have a constant dissociation factor. Fumaric acid has a greater dissociation constant than citric acid. As a result, two parts of fumaric acid in a fruit dessert will produce the same flavour effect as three parts of citric acid. The pH of a weak base may be likewise calculated, ammonia being a typical example. The formula will also identify a particular weak electrolyte in solution by reference to a table of dissociation constants. A few typical dissociation constants at 25°C of interest to food technologists are listed below.

Electrolyte	K
Ammonium hydroxide	1.8×10^{-5}
Benzoic acid	6.6×10^{-5}
Sulphurous acid	1.7×10^{-2}
Carbonic acid	3×10^{-7}
Boric acid	6.5×10^{-11}
Phenol	1.0×10^{-10}

It may also be shown that for a normal solution of a weak acid the degree of dissociation is equal to the square root of the dissociation constant. By using $M = \sqrt{K}$ the relative strengths of weak acids can be obtained from their dissociation constants.

Many factors influence the flavours of modern convenience foods. One undoubtedly is due to acidic components and this will be related to the acid's degree of ionisation and dissociation constant. A knowledge of pH is also valuable in estimating the shelf life of products by controlling the effects of oxidation and microbiological population level.

MOLAR SOLUTIONS

Normal solutions are regarded as the basis for volumetric analysis. They may be referred to as standard solutions since their strengths are known. A second standard solution often used in volumetric problems is the molar solution. A molar solution contains the molecular weight in grammes (that is, 1 mol) in 1 litre of solution. In practice, fractions or multiples of these solutions are employed.

The concentrations of normal and molar solutions may be the same. For instance, the equivalent of hydrochloric acid is 36.5, which is the molecular weight also. In the dibasic acid, sulphuric acid, the molecular weight is 98 and because of its dibasicity the equivalent is 49. The molar solution is thus double the strength of the normal

$\therefore$ 1 litre M H_2SO_4 is equivalent to 2 litres N H_2SO_4.

Both solution types are used in volumetric analysis because the weights of solute/m^3 are closely connected to their reacting weights, making calculations easier.

pH PROBLEMS

The acidity of any food system is more easily expressed as pH rather than hydrogen ion concentration for the reasons previously discussed. In pure water, the concentration of hydrogen ions and hydroxyl ions is equal at 22°C. Temperature changes above or below this figure gives other values.

The pH value of a solution may be defined as the logarithm to the base 10 of the reciprocal of the hydrogen ion concentration. A litre of pure water contains $\dfrac{1}{10\,000\,000}$ of a gramme ion of hydrogen. Therefore, the concentration of both the hydrogen and hydroxyl ions is $\dfrac{1}{10\,000\,000}$ of a gramme ion of each m^3. From the definition of pH it therefore follows that pure water has a pH of 7.

The mathematical product of both ions at 22°C is called the ionic product of water (K_w).

$$\therefore [H^+][OH^-] = K_w = 10^{-14}$$

The condition for neutrality will occur when both ions have the same concentration and this will only occur when each is 10^{-7}, since their product must always, whether they are equal or not, be 10^{-14}. The importance of K_w is that if either $[H^+]$ or $[OH^-]$ is known, the other can be found from the equation.

Example 1

What is the pH of centinormal $\dfrac{N}{100}$ sulphuric acid?

As a strong electrolyte in this dilution, only ions are present,

$$\therefore \ [H^+] = \frac{1}{100} = 10^{-2}$$

$$\therefore \ [OH^-] = \frac{[H^+][OH^-]}{[H^+]} = \frac{10^{-14}}{10^{-2}} = 10^{-12}$$

As pH is based upon hydrogen ion the solution has a pH of 2.

Example 2

What is the pH of centinormal $\dfrac{N}{100}$ sodium hydroxide?

Similarly, complete ionisation will occur from this strong alkaline solution.

$$[OH^-] = \frac{1}{100} = 10^{-2}$$

$$\therefore \ [H^+] = \frac{[H^+[OH^-]}{[OH^-]} = \frac{10^{-14}}{10^{-2}} = 10^{-12}$$

$$\therefore \ pH = 12.$$

Note: the additive pH is 14 and this will be a constant for all solutions involving variable hydrogen ion concentrations. It is, therefore, a relatively simple problem to calculate the pH of any strong acid or alkali if it undergoes complete dissociation into ions at a given concentration.

Ammonia and organic acids do not undergo complete dissociation in solution and therefore their pH cannot be calculated direct from K_w, but from a knowledge of their volume concentration and equilibrium constant. Three important formulae need to be noted: $M = \sqrt{(Kv)}$, $[H^+] = \sqrt{(K/v)}$ and $pH = -\frac{1}{2} \log K + \frac{1}{2} \log v$, where m = degree of dissociation, v = volume of solution 1 mol and K = dissociation constant. It will also be observed that $pH = -\log [H^+]$.

Example 1

Calculate the pH of a $\dfrac{N}{100}$ solution of benzoic acid ($K = 6.6 \times 10^{-5}$)

$$pH = -\tfrac{1}{2} \log K + \tfrac{1}{2} \log v$$
$$= -\tfrac{1}{2} \log 6.6 \times 10^{-5} + \tfrac{1}{2} \log 100$$
$$= -\tfrac{1}{2} \log (\bar{5}.8195) + \tfrac{1}{2} \text{ of } 2 = 1$$
$$= -\tfrac{1}{2}(-4.1805) + 1$$
$$= 2.0903 + 1 = 3.0903$$
$$\therefore pH = 3.0903$$

Example 2

What is the hydrogen ion concentration and pH of $\dfrac{N}{1000}$ ammonia solution?

$$(K = 1.85 \times 10^{-5})$$
$$[OH^-] = \sqrt{\frac{K}{v}} = \sqrt{\frac{1.85 \times 10^{-5}}{10^3}}$$
$$= \sqrt{(1.85 \times 10^{-8})} = 1.36 \times 10^{-4}$$

$$[H^+] = \frac{10^{-14}}{[OH^-]} = \frac{10^{-14}}{1.36 \times 10^{-4}}$$
$$= 7.353 \times 10^{-11} \text{ g ions/m}^3.$$
$$\therefore pH = -\log [H^+] = -\log (7.353 \times 10^{-11})$$
$$= -(\bar{11}.8665) = 10.1335.$$

Note: for practical purposes the pH would be reported to two decimal places.

The contraction and relaxation of animal muscle is a complicated biochemical process in which pH plays an important role. Lack of adenosine triphosphate (ATP) produces a rigid condition as adenosine diphosphate (ADP) is at a maximum. In the living animal there is sufficient energy and a fairly high pH to ensure that ATP can be produced for muscle relaxation. After slaughter, rigor rapidly sets in, the energy for this contraction coming from stored glycogen which is converted into lactic acid. The result is that the pH falls and this poisons the enzymes responsible for converting ADP to ATP.

Prior to slaughter beasts should be well rested in order to conserve their supplies of glycogen for the production of maximum lactic acid. Flavour, colour, texture and keeping quality depend upon a low pH. A fatigued animal will have used up most of its glycogen, and the oxidation products of carbon dioxide and water (carbonic acid) will produce a pH in the region of 6.7–6.9. At this pH micro-organisms responsible for food spoilage rapidly contaminate the meat, with resulting economic losses.

By using samples of pork and beef and adjusting the pH by injections of lactic acid and ammonia, a relationship between pH and tenderness can be inferred. When the samples are stored for 3–4 days at 0°C and then examined, it is found that toughness is at a maximum at a pH 5.0–6.0. The meat becomes progressively more tender at lower or higher pH levels. The variation in tenderness appears to be greater with beef than with pork.

The natural ageing of flour improves its baking quality with a resulting fall in pH. Chemical improvers like potassium bromate and ascorbic acid may be added to increase a flour's baking strength. No two improvers react in exactly the same way, but each will have an optimum pH activity. With potassium bromate a figure of 20 p.p.m. is desirable. Double this amount and the gluten will rot. Bromate analysis and pH control are therefore critical in producing good bread.

pH is important in fruit jelly stability. A pH of about 3.0 appears to suit most fruits. Syneresis or jelly weep may occur at a lower pH. However, some citrus pectins made into jelly show no appreciable syneresis at pH 2.0 or lower.

BUFFERS

A buffer solution is one which is capable of maintaining a constant pH value even when small amounts of acid or alkali may be introduced. For example, acetic acid may be added in small amounts to minimise microbiological spoilage in high moisture content confectionery such as fondant cream centres. Citric, malic, lactic and tartaric acids may be present as flavour components and also as an aid to the conversion of sucrose to invert sugar. The final taste will depend upon pH, which in turn is related to the acid dissociation constant, and whether it is buffered.

Buffer solutions are either a weak acid and one of its salts (usually the sodium salt), or a weak base and one of its salts. The essential principle is the weakness of the acid or the base in producing very few ions. In an acetic acid/sodium acetate system the following ions are present:

$$CH_3COOH \rightleftharpoons CH_3COO^- + H^+ \text{ (poor ionisation)}$$
$$CH_3COONa \rightleftharpoons CH_3COO^- + Na^+ \text{ (complete ionisation)}$$

The addition of a trace of acid, for example hydrochloric acid, will introduce additional hydrogen ions. These will combine with acetate ions from the buffer to produce acetic acid, which is relatively undissociated, so that the H^+ concentration is unaffected. Similarly, any alkali added would introduce hydroxide ions which would combine with hydrogen ions from the buffer to produce undissociated water. The result again is virtually no change in the pH. Buffer solutions may be regarded as solutions of reserve acidity and alkalinity.

Emulsion stability in sausages, fruit squashes, batters and doughs, etc. are all affected by pH changes. Any additive which can act as a buffer system is conducive to good quality.

SOLUBILITY PRODUCT

For practical purposes a large number of chemical compounds are insoluble in water. Examples include lead chloride and barium sulphate. However, the salts will have a very small degree of ionisation and in a saturated solution of barium sulphate there is a dynamic equilibrium between the solid salt and its ions:

$$BaSO_4 \rightleftharpoons Ba^{2+} + SO_4^{2-}$$
solid

Due to the very small solubility of the salt, the concentration of ions will be very small. At equilibrium the rate at which ions are produced will equal the rate at which they are recombined, so that a constant K_s will be produced.

$$K_s [Ba^{2+}] [SO_4^{2-}]$$

The equilibrium constant is the solubility product. It is defined as the product of the concentrations of the ions in the saturated solution of the salt. At 25°C the solubility product of barium sulphate $= 1.08 \times 10^{-10}$. If the solubility product is exceeded by the addition of more barium or sulphate ions, some barium sulphate is precipitated in order that the product remains a constant factor. Several applications in food problems arise as a result, among them the following.

1. Common salt may contain traces of calcium chloride and magnesium chloride which make the salt deliquescent. If hydrogen chloride gas is passed into a saturated, or near saturated, solution of brine, excess chloride ion combines with the sodium ion to produce pure sodium chloride. The salt may be packed, remaining free-flowing, unless over–exposed to high humidity.

2. The estimation of chloride ion in food mixtures, e.g. butter, is carried out by titration against standard silver nitrate solution, using potassium chromate as an internal indicator. The silver nitrate will react preferentially with chloride ions as it is easier to reach the solubility product of silver chloride than silver chromate. A dark red-brown precipitate of silver chromate indicates the disappearance of all chloride ions and hence the end-point of the reaction.

3. Routine qualitative inorganic analysis can detect the presence of metals in food known to have a potential health hazard. The detection of lead, cadmium and mercury depends upon their very small solubility products when precipitated as lead chloride, cadmium and mercuric sulphides, respectively.

PARTITION COEFFICIENT

Alcohol and water will mix together in all proportions whilst benzene and water, chloroform and water are immiscible for all practical purposes. Therefore, if a liquid or solid substance *C* which is soluble in each of two immiscible solvents, *A* and *B*, is shaken vigorously in both solvents until equilibrium is reached, the substance *C* will distribute itself in a definite ratio between the two solvents at a particular temperature. This constant is called the partition coefficient of the substance under test. It is found that if the same unit of concentration is used for both solutions, the partition coefficient is independent of this unit.

Example

10 cm³ of carbon disulphide and 10 cm³ of water were shaken with iodine at 18°C. The iodine content in the aqueous layer was determined volumetrically and the iodine in the organic layer was obtained by difference. Results were as follows:

1.68 g iodine in 10 cm³ of carbon disulphide.
0.0041 g iodine in 10 cm³ of water.

The partition coefficient of iodine between carbon disulphide and water is equal to the ratio:

$$K = \frac{\text{Concentration in carbon disulphide}}{\text{Concentration in water}} = \frac{1.68}{0.0041} = 410.$$

As a first principle it should be noted that the partition coefficient as defined above will only be constant if the solute *C* has the same molecular weight in both solvents. Association or dissociation should not occur.

The separation of chemical compounds using chromatographic techniques is discussed in detail at a later stage. However, it can be mentioned that the component may be regarded as a solute which partitions itself in a predetermined manner between two solvents. The rate of separation depends upon external factors, but the ratio of separation will follow the partition law. By using sophisticated methods the compounds may be identified on a partition coefficient basis.

A wider application is the extraction of organic substances using ether. This is the principle behind the Soxhlet extraction method. A food sample is wrapped in filter paper and placed in a special Soxhlet thimble so that the petroleum ether can be refluxed continuously over it. As the filter paper contains moisture the extracted fat separates according to its partition coefficient between ether and water. It is much more soluble in ether and hence a good separation occurs. The percentage of fat in flour, whole cream milk powder or biscuit can be obtained by this technique.

Amino acid separation in proteins depends upon the choice made by each amino acid between water and some suitable organic solvent, e.g.

butanol. In the washing of fatty substances with solvents for recovery purposes it can be shown mathematically that the best results are obtained using several small portions of extracting liquor rather than one or two large portions.

THERMOCHEMISTRY

Chemical reactions in which heat is given out are said to be exothermic and those in which heat is absorbed endothermic. The neutralisation of acid by caustic soda in the Kjeldahl estimation of food protein is highly exothermic. Great care is needed even after dilution in order to avoid serious burns. All exothermic actions give rise to positive solution heats as there is a rise in temperature.

In the preparation of batter using eggs the emulsion may break and curdled batter result unless the eggs are added at room temperature. Due to the negative heat of solution produced when sugar dissolves in the water from the egg the temperature falls, this increases surface tension and the fat and water phases may separate. In fondant preparation the heat produced as crystallisation occurs enables the mix to be worked more easily.

The kilogram-calorie (equal to a thousand small calories) is the unit of heat exchange and may be found either as C or kcal. Thermocalorific equations are similar to normal equations but with the amount of heat evolved shown as a plus (exothermic) or a minus (endothermic) heat change.

$$C + O_2 = CO_2 + 97°C$$
$$C + 2S = CS_2 - 25.4°C$$

One molecule of carbon dioxide is thus formed from its elements with the loss of 97 kcal to the atmosphere, its intrinsic energy, that is, the energy associated with its formation, is less by 97 kcal if the energy of all elements for reference purposes is assumed to be zero. Carbon disulphide is the exact opposite. It has absorbed heat from the atmosphere in its formation and has therefore an increased heat content compared to its elements.

The heat of formation of a compound is the heat change associated with the formation of a gramme molecule of the compound from specified states of its elements.

Hess's law of thermoneutrality or constant heat summation states that if a system A is converted into a system B the heat absorbed or evolved is independent of the method of passing from A to B. Carbon dioxide can be produced either direct from its elements, by action of heat upon a bicarbonate or by adding acid to either a carbonate or a bicarbonate. When all the necessary heat measurements are compiled it can be shown that to form one molecule of carbon dioxide requires the release of 97 kcal of energy.

Example

The heats of formation of hydrocarbon compounds can be deduced mathematically from experimental evidence.

Calculate the heat of formation of propane (C_3H_8), if its heat of combustion is 529 kcal. The heat of formation of carbon dioxide is 97 kcal and that of water 68.5 kcal.

It is required to find $3C + 4H_2 = C_3H_8 + x$ kcal.

$$C_3H_8 + 5O_2 = 3CO_2 + 4H_2O + 529 \text{ kcal}$$

$\therefore$ total heat evolved $= x + 529$ kcal.

By synthesis, the heat evolved in producing 3 molecules of carbon dioxide and 4 molecules of water is:

$$3C + 3O_2 = 3CO_2 + 291 \text{ kcal}$$
$$4H_2 + 2O_2 = 4H_2O + 274 \text{ kcal}$$

$\therefore$ total heat evolved $= 565$ kcal.

Using Hess's law, both heats can be equated to find x or the heat of formation of propane.

$$x + 529 = 565$$
$$\therefore x = 36 \text{ kcal.}$$

Note: the intrinsic energy or ΔH of propane is -36 kcal. All saturated hydrocarbons have a negative ΔH which indicates stability, as heat has been lost to the surroundings. Unsaturated compounds of carbon and hydrogen have a positive ΔH.

Acetylene is highly dangerous if attempts are made to liquefy it, due to its endothermic activity. The stability of compounds can therefore be related to their thermochemical activity.

The preparation of food additives may require a knowledge of the heat changes involved when reactants are mixed. By direct or indirect calculation techniques, the chemical engineer can design equipment of the right specification to withstand the heat exchanges involved.

EXERCISE 2

1. One molecule of ammonium carbamate (NH_2COONH_4) dissociates on heating into 2 molecules of ammonia and 1 molecule of carbon dioxide. The molecular weight of the undissociated compound is 78 and at 90°C it is 45 per cent dissociated. Calculate the vapour density at 90°C.
2. The percentage composition of a substance as determined by analysis is $C = 19.9$, $H = 6.9$, $N = 46.9$, $O = 26.3$. What is the empirical formula? Suggest a possible molecular formula if the substance is a waste product of animal metabolism.
3. A manufacturer of bleaching powder requires 100,000 litres of chlorine. How much common salt would he require as a source of his chlorine, allowing for 14 per cent loss during the process?
4. An aqueous solution contains 10^3 g of a permitted organic food colour/m^3. When 1 litre of the solution is treated with 100 cm³ of ether, 6 g of colour are extracted. How much more colour would be removed from the aqueous solution by:

    (*a*) a further 100 cm³ of ether;

    (*b*) two extractions each of 50 cm³ of ether?

5. How could Ostwald's dilution be used to identify individual food acids? Why is the law not obeyed by mineral acids?

6. Explain the terms pH and degree of ionisation. Given that the dissociation constant of propionic acid is 2×10^{-5}, calculate approximately for a $0.05N$ solution of propionic acid:

    (*a*) the degree of dissociation;

    (*b*) the hydrogen ion concentration;

    (*c*) the pH value.

7. Define the term solubility product. To what extent may the saponification of animal fat involve the concept of solubility product.

8. The heat of combustion of benzene (C_6H_6) is 783.4 kcal. Given that the heats of formation of carbon dioxide and liquid water are respectively 94.3 kcal and 68.4 kcal, calculate the heat of formation of benzene. What information about its stability could be deduced from its ΔH factor?

3 Metals in food

COPPER

The use of copper pans for heating food is desirable from the viewpoint of heat conductivity. It is an attractive metal, but it is expensive and housewives use aluminium containers wherever possible.

Copper's position in the electrochemical series indicates it to be less reactive than iron as regards attack by acids. This is a correct assumption, but temperature and time factors need to be considered. Apples which are boiled in copper vessels can produce sufficient copper ions for severe vomiting and diarrhoea to be caused in those who eat the product. Cases of poisoning have occurred where people have eaten bread, part of which was green. They presumed that the green was mould, when it was, in fact, caused by particles of bronze which had fallen into the bread from the baking machinery. The combined effects of oxygen and carbon dioxide can cause verdigris to be produced on a copper surface. Verdigris, being a basis copper carbonate, easily dissolves in fruit acids.

Copper poisoning of a chronic nature does not appear to exist, although the pure metal, if ingested, would irritate the alimentary canal, while soluble compounds cause only temporary inconvenience. As a trace element, to the extent of about 2 mg per day, it is needed for blood production and enzyme formation.

IRON

Iron pots are still used on the African continent as heating vessels for food preparation; however, the developed countries prefer steel vessels which are more durable and hygienic.

Iron is found in the haemoglobin of red blood cells and is stored as ferritin in the liver. Ferritin is an example of a conjugated protein containing in addition to iron, phosphate and protein.

A daily requirement of about 10 mg is recommended. Most of this comes from digested food, the remainder from water supplies and cooking utensils. In Europeans an excess of iron produces a condition known as siderosis, in which the skin assumes a brown colour and there appears to be a slight increase in blood pressure. The effects are less distressing than those due to copper, and in normal conditions a deficiency of iron rather than an excess is likely in our daily metabolism. Insufficient iron can result in anaemia. The mineral element is added to flour along with calcium to augment the food sources.

Figure 6 illustrates some of the sources of iron. It is interesting to observe that textured vegetable protein (T.V.P.) from soya bean is high in iron compared to beef.

Food	Iron p.p.m.	Food	Iron p.p.m.
Eggs	25	Haddock	10
Beef	37	T.V.P.	60
Brown bread	24	Cheddar cheese	6
White bread	18	Milk	1
Spinach	40	Apples	3

Fig. 6. Sources of iron in foods.

ALUMINIUM

Prior to the electrolysis of bauxite, the cost of preparation of aluminium from sodium precluded its use in food preparation. Modern production methods have brought the price down. Lightness, good heat conductivity and hygienic surface characteristics make it an ideal metal for domestic heat vessels.

Aluminium foil is used to protect a large range of foods. Even with the most sensitive foods, no odour or taste is imparted. It is impermeable and only a limited amount of gas diffusion will take place because of microscopic breaks in the foil. Aluminium will deter freezer burn in meat and is non-absorptive to liquids of all types.

There is no evidence to suggest that aluminium in the amounts in which it is likely to be consumed from using cooking utensils will produce any harmful effect. Indeed, the medical profession recommend its use as a therapeutic agent for gastric ulcers.

TIN

Canned foods are the obvious source of any tin that may find its way into the digestive system. A limit of about 250 p.p.m. is suggested, but this is on the grounds of food discoloration rather than toxicity. Tin itself is harmless. Food poisoning due to tinned foods can be traced to a rare infection with bacteria. It is, therefore, better to leave uneaten food in its tin rather than place it on a plate where it may be liable to fly infection, etc.

Solder, used for the side seam of a can, contains lead which is very dangerous in food. Fortunately, the electrolytic action between tin and lead produces a migration of tin into the food, leaving the lead on the can.

ZINC

The metal has a toxicity below copper and cases of poisoning are rare. Galvanised iron buckets which hold acid fruits, such as bilberries or rhubarb, are potentially harmful because zinc is high in the electro-chemical series.

New, galvanised iron pipes for conveying water have been known to cause trouble. On analysis, the zinc content was 40 mg/m³. A polythene pipe is the obvious answer to this problem. A recommended limit of 50 p.p.m. has been imposed. Unlike copper, which in very small amounts can catalyse milk rancidity, the presence of zinc at this level can be tolerated.

CALCIUM

Our earlier reading has indicated the paramount importance of this element in food. Apart from aiding bone structure, calcium ions are needed in the blood to allow Vitamin K to produce the prothrombin needed for blood clotting. A daily intake of at least 500 mg is recommended, and the milk industry is not slow to point this out in its attempts to encourage increases in milk consumption. The addition of chalk to flour still remains the most likely source of calcium intake.

Phytic acid in flour, which produces insoluble calcium phytate, reduces the natural intake from food. Fortunately, volumetric analysis can show the amount of added chalk needed to produce healthy growth. Milk protein or casein can combine with calcium to produce calcium caseinate, this being one factor which affects the reconstitution of dried milk powder. This is discussed in a later chapter on food preservation.

LEAD

Compounds of lead, like galena (lead sulphide), have been known for more than 2000 years, and since the extraction of the metal required no complicated technique it was used extensively for plumbing systems and the manufacture of cooking utensils. Lead acetate was used for 'sweetening' wine in medieval times, and today the use of it in illicit whisky appears to be on the increase in the U.S.A.

The use of lead as a liner for cider presses and tanks in Devon in the eighteenth century produced colic in the local population, a trouble which did not affect the rival cider county of Herefordshire where lead was not used.

Lead may escape detection as a potential health hazard if the concentrations are too small to give immediate discomfort. Several years may pass before the cumulative effects show themselves in the classical symptoms which include prolonged vague debility, convulsions and a blue line along the edge of the gums.

In Brest, a number of people were poisoned by bread made in ovens heated by lead-painted wood from old boats. Infants are sensitive to lead, and should be prevented from chewing yellow or red crayons coloured with lead chromate, or licking rain drops from exterior parts of houses painted with paint which contains lead. Dogs affected by lead show symptoms resembling distemper or hydrophobia.

Food itself is a source of lead which has been carried in solution from the soil media; 0.2 μg lead/kg of food is an average figure, but the concentration will vary considerably between different samples of types of foodstuffs. Lead arsenate is used as an insecticide on apples and pears, although organic insecticides appear to be taking its place. At present, the greatest pollutant source is the motor car where lead tetra-alkyls are used as anti-knock agents in petrols. Figure 7 shows the varying distributions in the atmosphere on a world basis.

Sampling point	Lead concentration μg/m^3
Detroit	4.8
Berlin (busy streets)	3.8
Berlin (quiet streets)	0.5
London (busy street)	3.2
London (hospital)	0.8
London (Blackwall tunnel rush hour)	23.0
Paris	4.8
Zurich	2.8
Remote area California	0.12
Boston (Summer tunnel)	44.5

Fig. 7. Atmosphere lead pollution.

Recent regulations have specified a limit of 2 p.p.m. for lead in food, with exceptions of about 5 p.p.m. in canned meat. Surveys conducted between 1966 and 1968 on food pesticide residues revealed that not one sample had lead in excess of the permitted limit. It would therefore appear that with suitable piping conditions for domestic water lead does not present a serious food problem. Atmospheric lead from industry and automobiles is a more difficult factor, and more research on this problem is indicated.

SODIUM AND POTASSIUM

The element sodiums, along with chlorine, is always found in body fluids. Indeed, common salt represents the main intake of sodium ions. As a flavouring substance salt is always present in a large number of foods; meat products and bread immediately spring to mind. There is no necessity to lay down a recommended allowance in a European diet unless the occupational activity produces excessive loss of salt. Manual workers in heavy industries restore the balance of sodium by consuming meat products and especially beer.

Potassium is similar to sodium chemically. Both elements assist in pH control. Blood has a pH close to 7.4, maintained by a buffer system involving a carbonate/bicarbonate system and a phosphate/protein relationship. Blood pressure is raised by the addition of salt to food. Special diets can be devised in which potassium chloride replaces sodium chloride as the flavouring component.

Sodium chloride, in combination with other curing agents, permits the thermal processing needed to produce pork luncheon meat to be lowered. 'Salt licks' for cattle are of paramount importance in hot climates if cramp and muscular fatigue are to be avoided.

The preservation of eggs using water glass, a viscous liquid containing sodium silicate, was a traditional method of extending their shelf life. When immersed the eggs were sealed externally against evaporation losses and bacterial invasion.

Baking powder will contribute its quota of sodium ions to food. An excess of bicarbonate has to be avoided on the grounds of colour and flavour problems, rather than because of nutritional needs.

Potassium is important as an essential element in grain cultivation and as a constituent of calcium phosphate in bone. It is also believed that potassium ions play a part in maintaining cell osmotic pressures. A figure of about 1000 mg/day is suggested for potassium intake.

Wherever possible sodium compounds are chosen before potassium compounds if both will provide the same physiological service. The choice is made on the grounds of economy, for potassium compounds are not as widely distributed.

EFFECTS ON FOOD

Certain metals, particularly those of variable valency such as cobalt, iron, copper and manganese, act as pro-oxidants in fat autoxidation by decomposing the unstable peroxide linkages with resulting free fatty acids. The processing and packaging of fat will require the elimination of any contamination with these metals, particularly copper. Storage in a cool, dark place will also inhibit rancidity tendencies. Tin, which as stated previously, is not regarded as a dangerous food contaminant, can affect the flavour of fats. Non-metallic surfaces of glass or lacquered tin are preferred.

The storage of meat in tin cans does not affect flavour unless the fat content is high. At temperatures below 0°C for a long period, white tin will turn into a more friable grey allotropic form. Tests on meat stored in this way for fifty years have produced edible meat with good flavour when the can was opened.

In any food process, and particularly in the preparation of food emulsifiers, the incorporation of sequestering agents may be needed to suppress the catalytic effect on flavour of metallic contaminants. A sequestering agent works by collecting metallic ions into a complicated molecule with suppressed ionisation tendencies. Sodium hexametaphosphate is useful in meat emulsions and citric acid in fruit squashes.

Free metal ions in soft drinks would alter the colour content as they catalyse the effect of light upon permitted edible colours.

It has been observed that when egg whites are beaten with copper-plated rotary egg beaters, they often develop a pink colour. This colour is produced by some constituents of the egg combining with traces of copper or iron. Fortunately, the coloration will disappear during baking, and the amount of metal involved is negligible.

The incorporation of walnuts and apples in a salad may produce an undesirable blackish-blue or purple colour. Slicing the apples with an iron knife causes the fruit enzymes in low pH conditions to produce ferric salts. When the ferric salt comes into contact with the tannin from the nut, and particularly with the skin of the fruit, the purplish colour develops. In nut-bread, walnuts may produce a dark colour.

Dark spots found in canned sweet potatoes are due to tannins which combine with the iron of the can. This discoloration will not occur without oxygen being present to oxidise ferrous ions to the ferric condition.

If asparagus is canned in an unlacquered tin a small amount of tin dissolves, producing a stannous salts which are good reducing agents and prevent the oxidation of the ferrous salts. When the water used in canning is high in iron content, sufficient tin may not dissolve to prevent the discoloration.

Brownish or black discoloration of vegetables in vinegar is caused by tannin from the vinegar combining with the iron of the food. Tannin-free vinegar eliminates the problem.

Chocolate ice-cream may contain greenish-black spots. These can be traced to the tannin supplied by the chocolate or cocoa and the iron from rusty spots in the can. In green-coloured beer, the iron was traced to exposed pipes, and the tannin from the hops.

Due to the perishable nature of milk, good hygienic conditions are essential. The milk is carried through glass or stainless steel vessels. Traces of copper are particularly dangerous. Milk is not well endowed with vitamin C, only about 20 p.p.m. occuring in cow's milk. Ascorbic acid (vitamin C) is easily destroyed by oxidation which is promoted by the presence of copper.

In the examples discussed, the presence of certain metals can be regarded as offensive to the aesthetic appreciation of food, rather than dangerous from a pathogenic viewpoint. Nevertheless, in cases involving vitamin retention, the control of contaminant is vital.

METAL TOXICITY

The metals which are most likely to cause disability, and even death, when present in food include mercury, lead, cadmium and arsenic. Arsenic may be regarded as a special case in that it exhibits both metallic and non-metallic properties and is also the most dangerous quantitatively.

Public interest has been focused on mercury as a contaminant in tuna fish. Mercury vapour has been recognised as dangerous but will normally only directly concern a very small percentage of the working population. Heavy industry uses mercury, which appears in the discharged effluent in the form of methyl mercury, and which in saline conditions finds its way into the fat portions of the tuna fish. Japanese fishermen living close to the effluent and whose main diet is fish other than tuna, developed the familiar symptoms of mercury poisoning. Brain atrophy and uncontrolled limb movement, similar to Parkinson's disease, rapidly ensued. Tuna fish, which is a recognised canned product in Europe, had its mercury content analysed. The level was higher than the previous natural mercury figure accepted as safe in normal food. U.S.A. sources indicated the presence of mercury in tuna fish in excess of 0.5 mg/kg. In England, a similar level in canned tuna was a distinct possibility. Further investigation showed that the intake of mercury from other food sources was less than 10 μg/day and that the mercury content of the major foods of our diet was extremely low and barely detectable. The average content in most canned fresh fish and shellfish is also low, but higher than in other foods.

It would appear that mercury poisoning is slight when viewed in perspective. Only those people who consistently eat large amounts of tuna fish, or fish from coastal areas with a high mercury content, appear to be at risk.

The publicity about mercury has undoubtedly affected tuna sales. Corned beef is another product the sales of which were reduced because of public concern about hygiene conditions in its preparation.

Apart from lead pollution in the atmosphere, a recent survey has been directed towards other possible sources of infection. Lead can occur in food and drink from natural sources, ranging from uptake of aquatic organisms from water containing lead, ingestion by farm animals and insecticidal sprays to manufacturing processes and pick-up from containers and equipment used in the preparation, cooking and storage of food.

The mean lead concentration in cans holding baby food is about 0.24 mg/kg which compares with only 0.04 mg/kg for baby food in jars. New regulations for foods specially prepared for infants and young children set a limit of 0.05 p.p.m. (mg/kg) for the lead content of such foods.

The lead content of cereals, meat, fish and vegetables was about 0.2 p.p.m. and milk only 0.03 p.p.m. Some shellfish varied between 1–2 p.p.m. However, there appears to be no great threat imposed from lead. The Committee on Food Additives estimates that a weakly intake of 3 mg lead from food and water for adults can be tolerated. The U.K. figure of 1.54 mg/week is well below this limit.

Cadmium, where used to plate containers, is rapidly dissolved by any acid food. Instances of such action have been reported from France when wine was left for only a few hours in such a container. In the U.K., tests have revealed that some shellfish may contain up to 4 mg/kg of cadmium. This is not unexpected as bivalves are able to concentrate heavy metals to a high degree.

Antimony may occur in the enamelling of cheap iron jugs or pails. In England, poisoning has resulted from drinking lemonade out of such containers. Nevertheless, neither cadmium nor antimony can be regarded as a serious threat to health in view of the small number of cases recorded.

Arsenic may be a natural contaminant of iron pyrites. In 1900, some 6000 people were severely poisoned and 70 died through drinking beer. Investigation showed that the beer contained arsenic derived from sugar refined with sulphuric acid obtained from pyrites that contained arsenic. (This danger does not arise today in the modern brewery.)

Arsenical sprays on crops are quite safe to man providing the exterior skins of the crops are well washed prior to processing.

The data on possible metal toxicity in food are considerable and there is no reason to suppose that with the sophisticated equipment now available any threat is posed to our national health. Statistically the chances of being involved in a road accident are at least a hundred times greater than accidental metallic food poisoning.

CORROSION AND TARNISHING

Corrosion is basically an electrochemical process. The term has become almost synonymous with iron although other metals show similar effects when subjected to the effects of moisture and oxygen. Iron will not rust in the presence of boiled water or when a drying agent such as calcium chloride is used to absorb moisture. As iron is above hydrogen in the electrochemical series it readily loses electrons to form positive ions, resulting in pitting of the metal surface and rusting occurs. In such circumstances any food cooked in an iron container would become discoloured and the presence of minute particles in the food would render it unacceptable.

The presence of carbon dioxide, sulphur dioxide and hydrogen sulphide will accelerate the corrosion effect by producing more ionic activity. Copper is 'weathered', producing a basic green-blue sulphate, which being water-soluble, poses the threats to food previously discussed. If copper vessels are used they should always be scrupulously cleaned after any food process involving heat.

In metal corrosion, the oxide is usually formed initially and the ease of oxide formation is greatest in the case of metals high in the electrochemical series.

Aluminium and magnesium are both above iron in the electrochemical series and therefore might be expected to corrode more rapidly than iron. In practice, this does not occur. The anomaly is due to the fact that a thin, but extremely durable, film of oxide is produced which is bonded firmly to the metal and prevents it from further attack. Iron does corrode more slowly but the layer of oxide formed easily flakes off exposing bare metal for more corrosive activity.

Aluminium and magnesium are examples of tarnishing as opposed to the more drastic effects produced by true corrosion. Silver is well below hydrogen in the series and has little attraction for oxygen. The blackening of silver is due to the tarnished effect produced by hydrogen sulphide in the presence of oxygen. Sodium, potassium and calcium will oxidise as soon as the free metal is exposed to air. The oxide stage will give way to the hydroxide and finally a layer of carbonate is produced.

The interior of aluminium pans should not be severely scoured after their use in food preparation. A mild treatment in neutral rather than alkaline conditions will remove scraps of food but leave the protective oxide layer intact. Zinc, tin and aluminium are all amphoteric in nature and respond to cleaning with neutral or slightly acid detergents.

ANALYSIS

The food analyst is concerned with those elements in food of a metallic nature that fall broadly into three classes. Class 1 comprises essential nutritive elements, typified by cobalt, copper, iron, zinc and manganese. Class 2 includes elements such as aluminium, chromium, nickel and tin which are regarded as non-nutritive and non-toxic. Class 3 covers the non-nutritive toxic elements, e.g. arsenic, antimony, cadmium, lead, mercury and selenium. These are known to have deleterious effects when present in food below 100 p.p.m.

Three main lines of analytical approach are available:
(*a*) gravimetric and volumetric;
(*b*) chromatographic; and
(*c*) spectrophotometric.

Tin may be estimated in the stannous (Sn^{2+}) condition by titration with standard iodine solution. It is necessary to prevent oxidation taking place as stannous solutions rapidly oxidise to the stannic (Sn^{4+}) condition. Usually about 20 g of sample is treated with a mixture of nitric and sulphuric acids and heated to oxidise away the organic matter. At a much later stage the combined effects of aluminium in the presence of strong hydrochloric acid reduce the tin to its lower valency and the colourless liquid is titrated in the presence of starch using centinormal iodine solution.

$$1 \text{ cm}^3 \ 0.01N \text{ iodine solution} \equiv 0.0005935 \text{ g Sn}.$$

Closely allied to the volumetric techniques for quantitative metal determination, in that colour changes are noted, is the colorimetric method used for copper. Copper is very sensitive to sodium diethyl-dithiocarbamate in alkaline solution, producing a yellow or brown coloration according to the metal concentration. By using a Lovibond Tintometer the colour can be matched visually to standard copper concentrations. This method depends upon accurate colour matching and requires experience, particularly when different hues are under observation. It has the merit of not involving a mathematical calculation.

In the analysis of organic complexes which contain a single metal, e.g. chlorophyll (magnesium) and myoglobin (iron), a gravimetric approach could be employed. A known weight of the sample is heated to 120°C for about 2 hours. This will remove the water. If the sample is now heated in a furnace for a further 4 hours at approximately 650°C, all the organic matter is removed leaving only the ash content. Using routine inorganic analysis, the iron may be estimated as one of its salts, and the iron content determined by calculation. The method requires considerable skill and is not recommended if a colorimetric estimation will give the information required. It is also time consuming and in a competitive world this increases the price of a product.

The separation of chemical compounds using chromatography has been known for a considerable time. However, the introduction of gas chromatography, as opposed to paper chromatography, only gained impetus after the Second World War. By using an inert carrier gas the sample is systematically broken down into its components on a partition coefficient basis. It is essentially a quantitative method.

Theoretically, anything that can be vaporised can be evaluated. The principle is that no two metal ions in a food sample will make the same separation choice in the gas chromatographic column. A recorder can be attached which will print off the metal concentration on to roll paper, the column height being indicative of metal concentration. Although the machine is very technical it can be operated by semi-skilled labour and is very accurate, with the human error content virtually eliminated.

By utilising the principles of the emission and absorption of light wavelengths, the use of a spectrophotometer in metal analysis has enabled rapid and accurate results to be compiled. Any desired part of the visible spectrum may be selected in the standard instruments. The relative amounts of radiant energy absorbed by a substance can be determined at different wavelengths and the appropriate curve plotted. From the curve the wavelength of the greatest absorption is selected. Using the wavelength, readings are obtained in a series of known concentration and a curve is plotted automatically. By reference to curves of known concentration the quantitative amount of any metal can be obtained.

The spectrophotometer can also be used to determine the rate of reaction of a chemical system, the nature of complex ions and their dissociation constants. A spectrophotometer is an expensive piece of equipment but is a good investment for routine metal checks. An example would be the arsenical content of food derived from plants due to the use of insecticidal sprays and dusts.

In milk analysis, the protein content can be determined in about 5 minutes as a direct reading using Amido black mixed with the milk.

After agitation, the sample is passed through a capillary tube and into a photo-electric cell.

Traces of copper which affect Vitamin C in foods can be readily detected by a spectrophotometer.

Every metallic ion will have its own characteristic print-out from a spectrophotometer. The result is analogous to human fingerprints. Although the machine is technically complex, its results can be easily interpreted and at a fast rate. Quality control, and its implications to food in general, depend to a large extent on equipment of the kind discussed, providing dependability and reliability.

EXERCISE 3

1. What metals would you regard as possible contaminants in bread? Indicate how modern bread production overcomes any problems that could arise.
2. Discuss the significance of employing stainless steel containers in milk processing plants. What laboratory tests for metal presence would you consider to be of importance if high quality is to be maintained?
3. Comment on the statement, 'iron is of paramount importance in meat and meat products'.
4. What factors influence the choice of metals to hold canned fruit?
5. A balanced diet should contain its proper mineral quota. What minerals would you consider essential? Briefly describe a quantitative test for calcium in flour.
6. Why is fat sensitive to metallic ions? What are the necessary conditions needed for good storage of either natural or manufactured fat products?
7. To what extent could metal toxicity represent a problem in confectionery goods? Suggest the practical steps needed in a quality control laboratory.
8. What advantages are offered by aluminium as a food container?
9. Describe the analytical equipment you feel is desirable to monitor the presence of metallic ions in:
    (a) meat products;
    (b) baby foods.

4 Organic chemistry

INTRODUCTION

Organic chemistry is particular to compounds of carbon, excluding simple molecules such as carbon dioxide, carbon disulphide, carbonates and bicarbonates. Carbon has the ability to combine with itself in long chain molecules and to have univalent atoms or groups attached to the carbon atoms. These aliphatic, or open chain compounds, include hydrogen, oxygen, nitrogen, sulphur or phosphorus in the molecule. Aromatic or cyclic compounds have the carbon atoms in a closed chain, again with different atoms attached to the valency bonds of the carbon atom. The number of carbon compounds known is enormous, but those which are concerned with food are of reasonable proportions.

In 1828, Wöhler heated an inorganic compound, ammonium cyanate, to produce urea. Urea is a waste product in animal metabolism. Prior to this discovery, it was thought to be impossible to produce in the laboratory any chemical found in a living organism. Perkin, in 1856, produced the world's first synthetic dye, mauveine. Today, colours and flavours for food requirements are the result of controlled chemical reactions. Unlike their natural counterparts, they are not affected adversely by pH, enzyme or oxidative processes when synthesised for a specialised function.

HOMOLOGOUS SERIES

Acetic acid is well known as a preservative compound. Vinegar is basically a 3 per cent solution with traces of other ingredients, including caramel, for colouring purposes in commercial vinegars. Mould inhibitors in bread can be based upon sodium propionate, produced from caustic soda and propionic acid. Both organic acids appear to exercise a similar food function. An examination of the chemical nature of North Sea gas reveals large amounts of the simplest organic compound known, namely methane. Campers and caravaners depend upon liquefied bottled gas, which is propane and butane. Methane, propane and butane appear to perform a similar function. All can be used in food preparation.

A close examination of the whole spectrum of organic compounds reveals that certain classes emerge. These classes all have a number of similar points in common, and are said to be members of an homologous series. A list of such properties could be summarised as follows:

(a) common formula;
(b) similar methods of preparation;
(c) physical properties;
(d) chemical properties;
(e) names, and
(f) they differ by CH_2 between successive members.

Example

Acetic and propionic acids are classified as monocarboxylic acids. The general formula for the series is $C_nH_{2+1}COOH$, the value for hydrogen being double the carbon plus one in the suffix part of the formula. All the acids in the series end in 'ic' and on further investigation the other properties show distinct relationships. Acetic acid, as the first member CH_3COOH, and propionic acid C_2H_5COOH as its higher homologue, are only two of the large family of monocarboxylic acids which occur either free, or in a combined form, in food products.

QUALITATIVE AND QUANTITATIVE ANALYSIS

The Lassaigne fusion test for the qualitative analysis of all organic compounds depends upon the ability of sodium, at a high temperature, to extract from the organic substance its elements as simple soluble inorganic salts.

About 0.1–0.2 g of the substance is heated in an ignition with a small piece of sodium. When all action appears to have ceased the tube is heated more strongly and the hot end of the tube is plunged into about 5 cms of distilled water contained in an evaporating basin, whereby the tube is broken and the filtrate is obtained. A few drops of ferrous sulphate solution are added to the filtrate. The mixture is boiled and then acidified with pure hydrochloric acid. If nitrogen is present in the original substance, a deep blue coloration or a precipitate is produced, either immediately or on the addition of a drop of ferric chloride solution. With sodium, the organic compound produces sodium cyanide which in the presence of ferrous sulphate forms ferrous hydroxide. Ultimately, ferric ferrocyanide or Prussian blue is produced.

$$Na + C + N = NaCN$$
$$6NaCN + Fe(OH)_2 = Na_4Fe(CN)_6 + 2NaOH$$

sodium
ferrocyanide

$$3Na_4Fe(CN)_6 + 4FeCl_3 = 12NaCl + Fe_4[Fe(CN)_6]_3$$

Prussian blue

Halogen detection follows a similar pattern except that the filtrate is acidified with nitric acid, followed by silver nitrate solution. The test depends upon the formation of the chloride, bromide or iodide of sodium which with silver nitrate produces characteristically coloured insoluble precipitates.

$$NaCl + AgNO_3 = NaNO_3 + AgCl$$
white

$$NaBr + AgNO_3 = NaNO_3 + AgBr$$
pale yellow

$$NaI + AgNO_3 = NaNO_3 + AgI$$
deep yellow

Sulphur, phosphorus and arsenic are detected by adding small quantities of the organic compound to strongly heated potassium carbonate containing about 5 per cent of sodium peroxide. The elements are thus oxidised to the appropriate acids.

$$S \rightarrow H_2SO_4, \quad P_4 \rightarrow H_3PO_4, \quad As \rightarrow H_3AsO_4$$

By applying routine inorganic analysis, the presence of the sulphate, phosphate and arsenate radicals can be proved. Sulphur can also be detected from the fusion test. A small filtrate portion with sodium nitroprusside solution gives a violet coloration.

Metals in organic compounds are detected by ashing the sample and then dissolving it in water. Group analysis test with inorganic reagents will then be applied to reveal metallic ions.

Quantitative analysis of an organic compound comprises one or more processes by means of which these elements are estimated. It is essential to have the substance under examination in a purified condition.

The following account of the methods used is only intended to indicate the general principles. Details of manipulation, quantities, etc. are dealt with in specialist texts.

Carbon and hydrogen are obtained by combustion analysis. A known weight of the organic compound is oxidised, and by calculation the elements are obtained from the weight of carbon dioxide and water produced. It is worth noting that there are no simple means of detecting the proportion of oxygen in an organic compound. By subtracting the percentages of all the other elements from 100 the oxygen value is obtained.

Nitrogen can be estimated either as nitrogen by Dumas's method, or as ammonia using Kjeldahl's technique.

In Dumas's method, the compound, when ignited with copper oxide, will produce carbon dioxide, water and nitrogen (or its oxides). By passing the gaseous products over heated copper oxide to decompose the oxides of nitrogen, and then collecting over potash, the carbon dioxide is absorbed. The residual gas is nitrogen. By measuring its volume from a known weight of substance (less than 1 g) and converting to S.T.P., the percentage of nitrogen can be determined.

Kjeldahl's method is more often used for estimating nitrogen in foods and fertilisers. The basis for the analysis depends upon the fact that when nitrogenous organic compounds are completely decomposed with hot, concentrated sulphuric acid, their nitrogen is obtained in the form of ammonium sulphate. Using distillation and volumetric analysis techniques (which are discussed in the next chapter in more depth) the nitrogen content of protein compounds is determined.

Chlorine, bromine and iodine are estimated by the method of Carius. The substance is oxidised with nitric acid at a high temperature in the presence of silver nitrate. Carbon dioxide and water are produced, but the halogens combine with silver to produce insoluble halides which are collected and weighed in the ordinary way.

Sulphur, phosphorus and arsenic are estimated as above without the addition of silver nitrate. The appropriate acids are produced, sulphuric, phosphoric and arsenic. Volumetric analysis will then reveal the amounts present and, by calculation, the percentage elemental composition.

As the bulk of food contains only carbon, hydrogen, oxygen and nitrogen, quantitative analysis is somewhat simplified and routine quality control is easily established.

ISOMERISM

Several cases are known of compounds which, although they have the same molecular formula, may have two or more different structural formulae. Each structural formula is an isomer of another form. Every isomer may or may not have a free existence. For example, it is theoretically possible to have 366 319 isomers from the hydrocarbon $C_{20}H_{42}$. On a more practical basis, there are two isomers for C_4H_{10}, three for C_5H_{12} and seven for C_7H_{16}.

Structural isomerism can be illustrated by *graphic* formulae, in which the valency bonds are shown as links to the carbon atoms. Alternatively, to save time and space, *rational* formulae may be employed. Figure 8 illustrates the structural and rational formulae of a number of items of food interest. It will be observed that the reactive groups involved can be distinguished in rational formulae.

All the compounds illustrated are said to be *saturated*, that is, each carbon atom is joined to a neighbouring carbon atom by a *single* valency bond. In other cases, a compound may contain two or more carbon atoms linked by a *double* or *triple* valency bond. These compounds are often highly reactive and have low melting or boiling points. They are said to be *unsaturated*. One important property of saturated compounds is that the hydrogen attached to a carbon atom can be substituted for an element or radical having the same valency. With un-

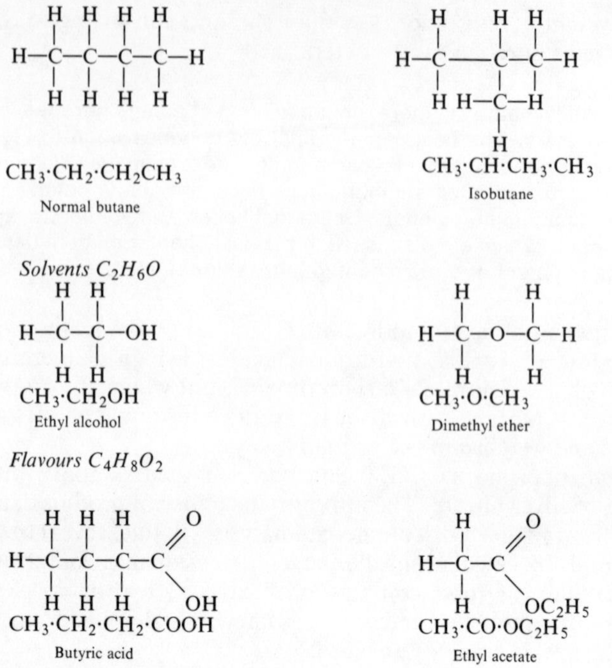

$$H-\underset{\underset{H}{|}}{\overset{\overset{H}{|}}{C}}-\underset{\underset{H}{|}}{\overset{\overset{H}{|}}{C}}-\underset{\underset{H}{|}}{\overset{\overset{H}{|}}{C}}-\underset{\underset{H}{|}}{\overset{\overset{H}{|}}{C}}-H$$

$CH_3 \cdot CH_2 \cdot CH_2 CH_3$

Normal butane

$$H-\underset{\underset{H}{|}}{\overset{\overset{H}{|}}{C}}-\underset{H-\underset{\underset{H}{|}}{\overset{\overset{H}{|}}{C}}-H}{\overset{\overset{H}{|}}{C}}-\underset{\underset{H}{|}}{\overset{\overset{H}{|}}{C}}-H$$

$CH_3 \cdot CH \cdot CH_3 \cdot CH_3$

Isobutane

Solvents C_2H_6O

$$H-\underset{\underset{H}{|}}{\overset{\overset{H}{|}}{C}}-\underset{\underset{H}{|}}{\overset{\overset{H}{|}}{C}}-OH$$

$CH_3 \cdot CH_2 OH$

Ethyl alcohol

$$H-\underset{\underset{H}{|}}{\overset{\overset{H}{|}}{C}}-O-\underset{\underset{H}{|}}{\overset{\overset{H}{|}}{C}}-H$$

$CH_3 \cdot O \cdot CH_3$

Dimethyl ether

Flavours $C_4H_8O_2$

$$H-\underset{\underset{H}{|}}{\overset{\overset{H}{|}}{C}}-\underset{\underset{H}{|}}{\overset{\overset{H}{|}}{C}}-\underset{\underset{H}{|}}{\overset{\overset{H}{|}}{C}}-\overset{\overset{O}{\|}}{C}\diagdown_{OH}$$

$CH_3 \cdot CH_2 \cdot CH_2 \cdot COOH$

Butyric acid

$$H-\underset{\underset{H}{|}}{\overset{\overset{H}{|}}{C}}-\overset{\overset{O}{\|}}{C}\diagdown_{OC_2H_5}$$

$CH_3 \cdot CO \cdot OC_2H_5$

Ethyl acetate

Fig. 8. Structural and rational formulae.

saturated compounds this is not the case; the entering radicals or groups are attached by the process of addition.

Other cases of isomerism occur in which the isomers cannot be illustrated using graphic or structural formulae. The term steroisomerism is used and two distinct types are recognised: (*a*) optical; and (*b*) geometrical.

Tartaric acid has the graphic or rational formula.

$$\begin{array}{c} CH(OH).COOH \\ | \\ CH(OH).COOH \end{array}$$

In the purification of the acid from grapejuice a second acid was isolated. The acid was called racemic acid and appeared identical to tartaric acid in every way except that in solution it had no effect on polarised light. Tartaric acid is strongly dextrorotatory. Pasteur examined very carefully the crystalline nature of the sodium salts of both acids. Racemic acid had equal quantities of crystals whose facets were the mirror image of each other. They were related as are a pair of hands. It is impossible to superimpose a left hand on a right hand, for the thumbs and fingers will not fit. With tartaric acid, the crystals are only of the right

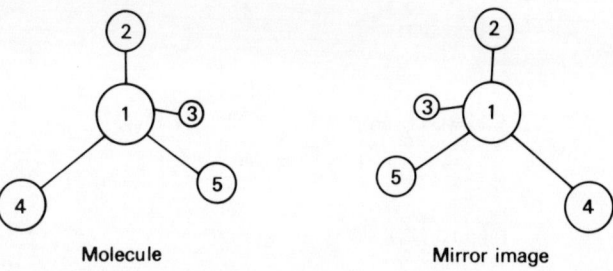

Fig. 9. Optical isomerism.

hand type. Figure 9 illustrates in a general way the appearance of an optically active compound if it could be viewed in a mirror.

Other food examples include amyl alcohol, lactic acid and proteins. These compounds have at least one carbon atom (asymmetric) which is linked to four different atoms or groups.

Maleic acid is important as one of the constituents in alkyd resins. These form the basis of enamels, lacquers and varnishes. Fumaric acid has the same molecular and structural formula and has a pleasant fruit acid flavour.

$$H—C—COOH$$
$$\|$$
$$H—C—COOH$$

Maleic (cis)

$$H—C—COOH$$
$$\|$$
$$COOH—C—H$$

Fumaric (trans)

A closer examination reveals that maleic acid has both the hydrogen and the carboxylic atoms on the *same* side of the carbon atoms, whilst in fumaric acid identical atoms or groups are *across* the double bond joining the carbon atoms. Both compounds exhibit geometrical isomerism which is characterised in compounds showing unsaturation. The double bond prevents free rotation and therefore the formula of each compound, although structurally identical with the other, cannot be superimposed. To distinguish each isomer the terms 'cis' and 'trans' are used.

Geometrical isomerism occurs in fat chemistry. Oleic acid is a constituent of many oils and edible fats. It has one double bond and is therefore unsaturated. It is necessary for certain biochemical processes and is known as an essential fatty acid (E.F.A.). As it melts at 14°C, the free acid is normally liquid at room temperatures. The geometrical isomer of oleic is elaidic acid. With a melting point of 52°C, it is always a solid at normal temperatures. Animals fed with elaidic acid in place of the normal oleic acid are liable to lose weight, the coat becomes dull and the animal is listless.

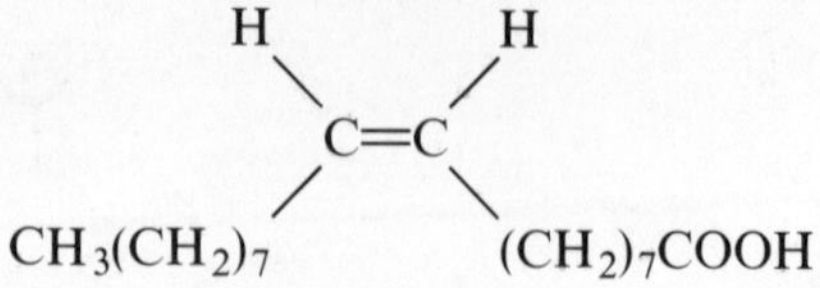

$$\begin{array}{ccc} H & & H \\ \backslash & & / \\ & C{=}C \\ / & & \backslash \\ CH_3(CH_2)_7 & & (CH_2)_7COOH \end{array}$$

Oleic acid (cis)

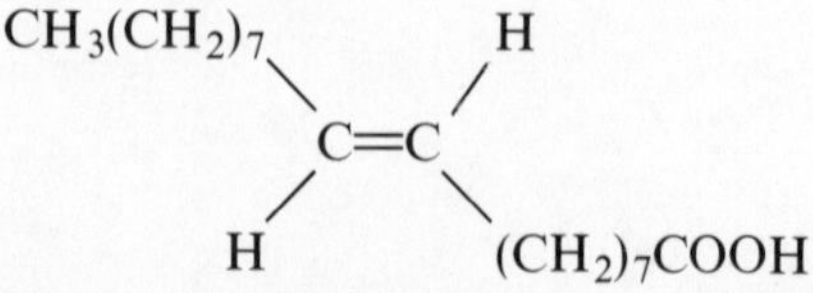

$$\begin{array}{ccc} CH_3(CH_2)_7 & & H \\ \backslash & & / \\ & C{=}C \\ / & & \backslash \\ H & & (CH_2)_7COOH \end{array}$$

Elaidic acid (trans)

Both acids have the same molecular formula $C_{17}H_{33}COOH$ and the body chemistry is confused. Elaidic acid is absorbed into building an E.F.A. molecule. The imperfect molecule behaves as an anti-E.F.A. factor and prevents the absorption of essential oleic acid.

It can be shown that if a fatty acid contains n double bonds it will have $2n$ possible isomers. Linoleic acid has two double bonds and consequently four isomers.

HYDROCARBONS

These are the simplest organic compounds and are represented by: (*a*) alkanes or paraffins; (*b*) alkenes or olefins; and (*c*) benzenes or aromatic.

The alkanes form an homologous series of general formula $C_nH_{2n}+2$. Methane CH_4, ethane C_2H_6, propane C_3H_8 and butane C_4H_{10} are its first four members. As fuel gases they have some importance in food but their main virtue is their ability to undergo *substitution*. With chlorine, methane will produce four new compounds having quite different properties from the parent hydrocarbon.

$$CH_4 + Cl_2 = CH_3Cl + HCl$$
methyl
chloride

$$CH_3Cl + Cl_2 = CH_2Cl_2 + HCl$$
methylene
dichloride

$$CH_2Cl_2 + Cl_2 = CHCl_3 + HCl$$
chloroform

$$CHCl_3 + Cl_2 = CCl_4 + HCl$$
carbon
tetrachloride

Chloroform is used as a fat solvent in determining the hardness of fats. It replaces carbon tetrachloride which is carcinogenic.

Alkenes having a double bond contain less hydrogen than the saturated alkanes and are *unsaturated* hydrocarbons. Their general formula is C_nH_{2n} and the simplest member is ethylene C_2H_4. Structurally, it can be represented thus:

$$
\begin{array}{ccc}
H & & H \\
\diagdown & & \diagup \\
& C\!=\!C & \\
\diagup & & \diagdown \\
H & & H
\end{array}
$$

The homologues of ethylene include propylene C_3H_6, butylene C_4H_8 and amylene C_5H_{10}, etc. In the alkenes, the possibility of isomerism is greater than with the alkanes due to variation in the arrangement of carbon atoms and because the double bond may be in one of several positions.

Contrary to what might be expected, the presence of the double bond or unsaturation point is a sign of weakness, not strength. This is borne out by comparing the heats of formation of alkanes and alkenes.

Alkenes are more reactive than alkanes and easily undergo *additive* reactions. The addition of two univalent atoms breaks the double bond and produces saturation.

$$CH_2\!:\!CH_2 + H_2 = CH_3.CH_3$$
$$CH_2\!:\!CH_2 + Cl_2 = CH_2Cl.CH_2Cl$$

Hydrogen can be added on to the double bond using high temperature, pressure and a nickel catalyst. The process known as *hydrogenation* will convert oils into fats. The process is of great economical importance as it is involved in changing whale oil or other edible oils into margarine.

Ethylene will also polymerise to produce a plastic material polyethylene, more often called polythene. Depending upon the number of basic ethylene units condensed in the polymer, the result can range from waxy solids to the familiar, tough, clear resinous material. Polythene is ideal for packaging food; it is hygienic and will not encourage bacterial growth. It has been used as a liner for fibreboard holding liquid egg. Polypropylene is an important laminate for packing foods such as rice, dried vegetables and mixed fruits.

A curious use for ethylene is in the ripening of green fruit. The introduction of small amounts of ethylene into a room holding unripe unwrapped bananas will produce a ripe crop in a matter of hours as opposed to days.

Benzene C_6H_6 is the type member of the aromatic hydrocarbon series. It combined both the alkane and alkene features as it consists of alternate double and single bonds. In order to save time, it is customary to omit the symbol sum of carbon and hydrogen atoms. It has a molecular formula C_6H_6 and is expressed in the form of equations as

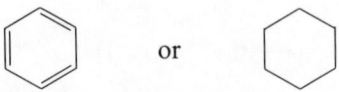

or

At each point of the regular hexagon, atoms or groups can replace hydrogen to produce *intermediates*. Toluene or methyl benzene is an example.

CH_3

The benzene ring is always a sign of an aromatic compound. Like alkanes and alkenes, the benzenes form an homologous series of general formula C_nH_{2n-6}. As a class, they are less reactive than the alkenes and, despite the three double bonds, prefer to undergo substitution rather than addition reactions.

Other ring systems are known which contain elements other than carbon. Nitrogen may be present, or sulphur. In these cases the molecule is said to be *heterocyclic* rather than *homocyclic*.

ALCOHOLS

The term alcohol is applied to organic compounds which contain one or more hydroxyl groups. Three classes of alcohols can be recognised depending upon how the OH group is attached to carbon. In a primary alcohol, the OH group is attached to the last carbon atom of the chain; the group is therefore CH_2OH. In secondary alcohols, the OH group is attached to a carbon atom linked by two valency bonds to other carbon atoms; the group is therefore $\diagdown$ CHOH. Finally, a tertiary alcohol has an $\diagup$ OH group linked to a carbon atom of which the three remaining bonds are linked to other carbon atoms; the group is therefore $-$COH.

Butyl alcohol C_4H_9OH shows all three types:

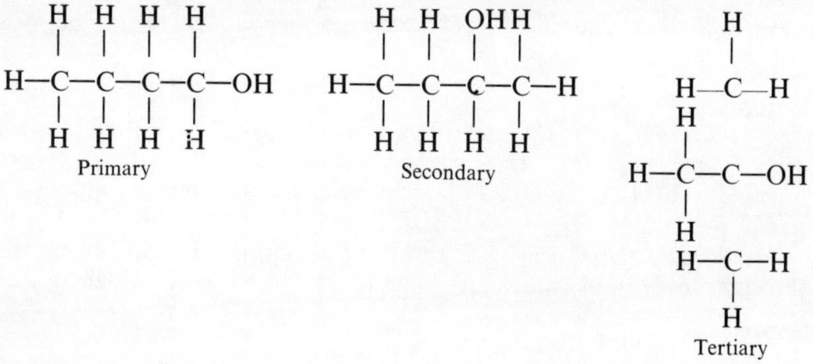

Primary   Secondary   Tertiary

The three classes of alcohols can be distinguished by their different oxidation products.

Primary   ⟶ Fatty acid
Secondary ⟶ Ketones
Tertiary   ⟶ Complex mixture of acids, ketones, etc.

Ethyl alcohol C_2H_5OH is the most important alcohol. It is a product of the fermentation of glucose with the enzyme zymase. All alcoholic beverages contain it.

Ethyl alcohol will react with acetic acid to form ethyl acetate. This compound is known as an ester. Esters are important as flavour components. Methyl alcohol is toxic in spirits but esters can be produced from it to advantage. For example, methyl butryate produces a pineapple odour. Although ethyl alcohol and isopropyl alcohol would act as good solvents for volatile flavour constituents, manufacturers of flavours may choose propylene glycol which has a much higher boiling point and ensures a better separation. Isopropyl alcohol is preferred in the U.K. for essences, whereas on the Continent the more expensive ethyl alcohol (ethanol) is used.

Off flavours can be traced to esters found in fat which has decomposed to glycerol and free fatty acids. Glycerol is a trihydric alcohol, with the following composition:

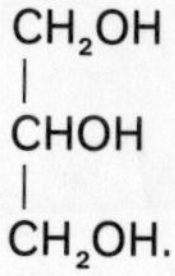

$$CH_2OH$$
$$|$$
$$CHOH$$
$$|$$
$$CH_2OH.$$

When overheated it breaks down to produce acrolein. This affects the eyes and is associated with the overheating of fat in producing chipped potatoes. Glycerol is useful to the baker as a basis for producing G.M.S. (glyceryl monostearate), an important emulsifying agent.

Advanced food science

ACIDS

All food acids contain the carboxylic group $-C\begin{smallmatrix} \nearrow O \\ \searrow OH \end{smallmatrix}$ as a functional feature.

Fatty acids, which are an integral part of fat molecules, contain only one carboxylic group and are, therefore, monobasic acids. Their general formula is $C_nH_{2n+1}COOH$ and acetic acid may be taken as representative of the group.

In all cases they form as a result of oxidation, the alcohol passing through an unstable aldehyde stage to the stable acid. Aldehydes are characterised by the group $-C\begin{smallmatrix} \nearrow H \\ \searrow O \end{smallmatrix}$ and are generally recognised as unpleasant compounds in food products if they are of the aliphatic type. The aromatic aldehyde found in essential oil of almonds in benzaldehyde C_6H_5CHO. Unless the oil is kept in small and stoppered containers it will rapidly oxidise, producing flavourless benzoic acid.

The dicarboxylic acids contain two COOH groups/molecule and the series contains a number of members of interest to the food industry. Other cases occur of acids containing a hydroxyl group. This group will tend to reduce the sourness of the lower members of the acid series. Lactic acid is well known in milk and milk products. It combines a sharp flavour with sweet overtones. Citric acid is a tricarboxylic acid, and bitter lemon owes much of its flavour to this component. Benzoic acid is an important aromatic acid effective in controlling the growth of micro-organisms in food of low pH.

Figure 10 shows some of the more important food acids and their applications.

Name	Rational formula	Type	Application
Acetic	$CH_3 \cdot COOH$	saturated	preservative
Propionic	$C_2H_5 \cdot COOH$	saturated	mould inhibitor
Butyric	$C_3H_7 \cdot COOH$	saturated	fat rancidity
Lauric	$C_{11}H_{23} \cdot COOH$	saturated	coconut oil
Stearic	$C_{17}H_{35}COOH$	saturated	animal fat
Oleic	$C_{17}H_{33}COOH$	unsaturated	palm oil
Lactic	$CH_3CH(OH) \cdot COOH$	saturated	yoghourt
Oxalic	$COOH \cdot COOH$	saturated	strawberry
Malic	$CHOH \cdot COOH \cdot CH_2 \cdot COOH$	saturated	rhubarb
Tartaric	$(CHOH)_2 \cdot (COOH)_2$	saturated	baking powder
Succinic	$(CH_2)_2 \cdot (COOH)_2$	saturated	dough
Citric	$C(OH)COOH \cdot (CH_2COOH)_2$	saturated	citrus fruit

Fig. 10.   Important food acids.

AMINES

These may be regarded as substituted derivatives of ammonia, and, like it, are strongly basic. Amines are distinguished as primary, secondary or tertiary depending upon whether 1, 2 or 3 atoms of hydrogen in ammonia have been displaced by alkyl groups.

Ethylamine $C_2H_5NH_2$ Diethylamine $(C_2H_5)_2NH$
Primary   Secondary

Triethylamine $(C_2H_5)_3N$
Tertiary

All proteins contain the amino group $-N\begin{smallmatrix} H \\ \diagup \\ \diagdown \\ H \end{smallmatrix}$, and when proteins decompose, the formation of simpler products results in the release of amine compounds. Ptomaines, which are produced in conditions of putrefaction, result from the decarboxylation of amino acids, for example.

$$CH_3.CH.NH_2.COOH \longrightarrow CH_3CH_2NH_2 + CO_2.$$
alanine   ethylamine

Proteolysis of meat protein produces poor texture and amine release. The rise in pH will affect colour and odour. The characteristic smell of bad fish is due to the breakdown of a complex nitrogenous compound, trimethylamine oxide. Bacteria break it down in the fish after death to the tertiary amine, trimethylamine.

ESTERS

Alcohols will react with acids to produce esters. The action is reversible, but if strong sulphuric acid is added the water produced is absorbed and hydrolysis is thereby prevented.

$$C_2H_5OH + HCl \rightleftharpoons C_2H_5Cl + H_2O$$
$$C_2H_5OH + CH_3COOH \rightleftharpoons CH_3COOC_2H_5 + H_2O$$

If the acid is inorganic, the product is often referred to as an *alkyl halide*, the term ester in food chemistry meaning that an organic acid is involved.

Esters formed from simple alcohols are pleasant-smelling liquids and can be found in suitable blending proportions along with pure chemicals as synthetic flavouring agents. Glycerol can combine with either the same fatty acid to produce simple triglycerides or, if more than one kind of fatty acid is involved, a mixed triglyceride can result. The significance of these reactions is dealt with in the chapter dealing with fats.

NITRILES

These are equivalents of inorganic cyanides since they contain the group $-C \equiv N$. Nitriles are often named after the acid they produce on hydrolysis:

$$CH_3.CN \xrightarrow{\text{hydrolysis}} CH_3COOH.$$

acetonitrile   acetic acid

Nitriles are dangerous in food. They must be removed if it is suspected they may be present as a result of biochemical activity. Manioc, which is a high calorie root crop in the tropics, must be well washed to remove nitrile traces from it. In the presence of any free mineral acid, the resulting ammonium salt is rapidly acted upon to produce the free organic acid.

$$C_2H_5CN + 2H_2O \longrightarrow C_2H_5COONH_4$$

propionitrile   ammonium propionate

$$C_2H_5COONH_4 + HCl \longrightarrow C_2H_5COOH + NH_4Cl$$

propionic acid

IMIDE

This group resembles the amines but has one hydrogen atom less than a primary amine and is not joined to alkyl groups. The imide group can be designated as $\diagdown NH \diagup$ showing that it has two free valency arms.

Saccharin is a well-known sweetening agent, it has approximately 400 times the sweetness of an equivalent weight of cane sugar. However, unlike sugar, it provides the body with no nutritional requirements and is not metabolised. It would be difficult to incorporate it into a food mix and it has the disadvantage of imparting a bitter aftertaste to baked goods. The chemical name for it is ortho sulphobenzimide and the starting raw material is toluene.

Chemically, its formula is relatively simple:

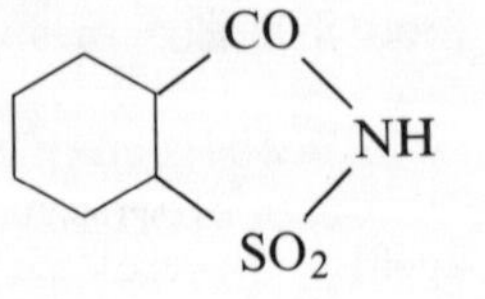

As a sweetener in foods for diabetics it has a useful function, and it is a support for the 'weight-watchers' diet.

AMIDE AND AZO GROUPS

The amide group $-CO.NH_2$ is associated with katabolic processes. When ammonium acetate loses a molecule of water, acetamide results. The pure compound has only a faint odour, but as usually prepared, it has a strong smell of mice.

Urea or carbamide $NH_2.CO.NH_2$ is a white powder excreted in urine at the rate of about 25 g/day. Its uses to the food industry are threefold. First, as a valuable source of nitrogen it is an ideal fertiliser, and secondly, it is used in the preparation of urea–formaldehyde resins which can produce a good clean surface for food preparation. In recent years, its third use has been to incorporate it into the feeds of cattle and sheep. They are able to absorb it and use it in the synthesis of protein materials. Approximately one-third of the normal protein fed to these animals can be supplied by urea. It has helped to conserve true protein supplies and assist in economic production of first class meat.

An azo group $-N:N-$ is found in synthetic organic colours suitable for incorporation in food. When an aromatic primary amine is treated with nitrous acid below 5°C, an unstable compound called a diazonium salt is produced. This diazotisation process is the intermediate link between initial and final colour development. If the diazonium salt is now reacted with an aromatic alcohol (phenol), the colour is obtained by a process known as coupling.

The following equations show the sequence of events.

aniline $+HCl \xrightarrow{\text{addition}}$ aniline hydrochloride

$NH_2.HCl + O:N.OH \xrightarrow{\text{below 5°C}} N:N.Cl + 2H_2O$

nitrous acid

benzene diazonium chloride (diazonium salt)

$N:N.Cl + $ phenol (hydroxy benzene) $\xrightarrow{\text{alkaline condition}}$ $N=N$ $OH + HCl$

hydroxyazobenzene (yellow-brown dye)

Further reference to this process will be made in Chapter 6. No attempt should be made at this stage to learn the above equations, the significance of the process in being able to produce edible food colours is the important feature.

There are several other important functional groups in organic chemistry which have not been discussed in this chapter. Reference will be made to them as and when they occur, e.g. the solubilising effect of the sulphonic acid group in complex organic molecules.

EXERCISE 4

1. What is meant by substitution? Write equations, and suggest possible names, for the products obtained by the substitution of chlorine in ethane.
2. Suggest three chemical tests to prove an unknown liquid is acetic acid. Illustrate the use of the acid in food in general.
3. How would you recognise a given liquid as being an ester? What differences exist between the esters of monohydric alcohols and those produced from glycerol, from a food point of view?
4. What chemical tests could distinguish:
   (a) ethylene from ethane;
   (b) butyric acid from lactic acid;
   (c) methylamine from ammonia?
   Discuss briefly the significance of any *three* compounds in food technology.
5. 'The alcoholic group is of supreme importance to the food technologist.' Comment on this statement and give *two* more reactive groups of value in the understanding of food processes.
6. Write an account of isomerism with particular reference to food examples.
7. A food product contains an organic acid. After its isolation and purification, 15.5 cm^3 of normal sodium hydroxide solution were required to neutralise 0.9145 g of the acid. calculate the acid's molecular weight, indicate a possible name and structural formula and suggest the possible food source.
8. Explain the term homologous series with reference to alcohols and acids of food interest.

5 Physical and chemical aspects of nutrients

CARBOHYDRATE

All carbohydrates represent energy sources for the body. Starch and cellulose are polymers made up from the simple sugar unit of glucose. Cereal foods, barley, flour and potatoes are rich in carbohydrate, providing starch which can be broken down by enzymes in the human body.

Both starch and cellulose are isomers and yet cellulose is useless as a human food supply unless subjected to vigorous chemical treatment. Both compounds have a slightly different structural formula and cannot be superimposed. In starch, the individual glucose units are joined by 'oxygen bridges' which are all in the same position. The cellulose molecule has an alternative bridging system in which the units are linked by 'oxygen bridges' which are alternately upwards and downwards. In the presence of the enzyme cellulose, starch is unaffected but cellulose is rapidly broken down to smaller absorbable units. By comparison, the diastatic enzymes found in malt readily break down starch to soluble sugars. Each enzyme may be regarded as a 'key' which fits only one 'lock', the substrate starch or cellulose.

It is, therefore, apparent that the structural formula of a complex molecule is more important than the molecular formula if differences between them have to be explained. Starch is a great deal more complex than sugar, but far less intricate than food proteins where the polymer chains may consist of a large number of different amino acid units. The presence of certain reactive groups such as aldehyde (CHO) and secondary alcohol groups (CHOH) account for the different physical and chemical properties of the carbohydrates.

Starch is used in cookery as a thickener for soups and sauces. It possesses marked adhesive qualities when hydrated and can be used as a binding material for the framework of biscuits, cakes and bread. These characteristics are a feature of its gel-forming properties.

Cereals provide the major starch source. The granule size and shape found are peculiar to each type of plant; microscopic examination readily identifies each sample. At room temperature the starch granule is insoluble in water. There is no surrounding membrane; the contents of the swollen granule are held together by hydrogen bonds. Variations in humidity affect the hygroscopic nature of starch.

When flour absorbs water the heat of hydration decreases with increased moisture content, although the graph is not linear. It is identical for large and small granules of wheat and potato starches.

As starch granules are heated in water there is a large volume and viscosity increase. Translucence occurs and solubility increases. The process is irreversible and the granules cannot be obtained in their original form. The term gelatinisation covers these changes. It has been found that the temperature at which swelling begins is variable for the sample under test and that the range of temperature over which the swelling occurs, and the type of gel formed, are characteristic of each kind of starch. For example, potato and tapioca have a similar gelation range which begins at about 65°C. Cornstarch is able to gelatinise over a wide range of temperature.

Swelling is always greater in distilled water than in salt solutions. Certain salts may lower the gelatinisation temperature to such an extent that the swelling process can be observed at 20°C. In general, swelling increases with an increase in concentration of the salt used, and differences occur with the anions involved, for example OH has more effect than Cl. Normally the longer a dough remains in a plastic condition the greater will be the volume of the baked loaf.

A number of factors contribute to the stiffness of the gel. These include (*a*) the size of the starch molecule; (*b*) the shaking or stirring time during gelatinisation; (*c*) the pH; (*d*) the enzyme hydrolysis; (*e*) the dry heat; (*f*) the temperature; and (*g*) the fat content. A relationship exists between gelatinisation and viscosity; swelling increases granule size and thus the viscosity of the resulting paste. Viscosity measurements are done on the cooled gel. Dough is a most complex rheological material and this is directly related to the complexity of the starch molecule.

Glucose and fructose are isomeric simple sugars which have different sweetening effects. Fructose is much sweeter and this is reflected in its formula.

$$CH_2OH(CHOH)_3CHOH \quad CHO \qquad CH_2OH(CHOH)_3CO \quad CH_2OH$$

glucose (dextrose)   fructose (levulose)

The presence of a *ketone* group ($\diagdown C{=}O$) in fructose implies a pleasanter flavour than glucose which contains an unpleasant aldehyde group.

Sucrose is used in far larger quantities than any other monosaccharide or disaccharide sugar as a sweetening agent. Dextrose is found in ice-cream, preserves, canned foods and in confectionary. Maltose and lactose in quantity are used in areas of confectionary work. Relative sweetness of sugars is affected by temperature as well as concentration, the intensity of sweetness being increased as the temperature is raised.

Weak acids, used in confectionary work, lemon juice, fruit juice or cream of tartar all cause sugar inversion. Fondant hydrolysis by these agents occurs at a slow rate during storage. Cream of tartar can impart a slightly sour taste to the fondant, which to some people is regarded as a desirable flavour impact. Chocolate cream centres contain liquefied fondant. This liquefaction is due to the enzyme invertase which is added to the fondant when it is melted for moulding. Invertase can also be used for bon-bons which dry out more rapidly than chocolate-coated candies, although a small amount of egg white added to fondant minimises the evaporation.

To improve the free-flowing properties of white sugars the optimum storage conditions appear to be about 38°C and a relative humidity of less than 60 per cent. Brown sugar contains about 4 per cent more inherent moisture than white sugar and may be stored successfully at 24°C and a humidity of 60–70 per cent.

FAT

Glycerol as a trihydric alcohol combines with three molecules of fatty acid to produce a glyceryl ester. Depending upon the physical nature of the resulting triglycerides, an oil or fat molecule may result. The greater the degree of unsaturation of the fatty acid, the less stable the molecule becomes to oxidative processes. A typical fat molecule is tristearin and an oil molecule would be presented by triolein.

$$CH_2OH \qquad\qquad\qquad CH_2OCOC_{17}H_{35}$$
$$CHOH \ +3C_{17}H_{35}COOH \rightleftharpoons CHOCOC_{17}H_{35} \ +3H_2O$$
$$CH_2OH \qquad\qquad\qquad CH_2OCOC_{17}H_{35}$$

glycerol	stearic acid	tristearin
(one mole)	(three moles)	(simple triglyceride)

$$CH_2OH \qquad\qquad\qquad CH_2OCOC_{17}H_{33}$$
$$CHOH \ +3C_{17}H_{33}COOH \rightleftharpoons CHOCOC_{17}H_{33} \ +3H_2O$$
$$CH_2OH \qquad\qquad\qquad CH_2OCOC_{17}H_{33}$$

glycerol	oleic	triolein
(one mole)	(three moles)	(simple triglyceride)

Natural oils tend to be mixed triglycerides, each molecule containing more than one type of fatty acid. All oils and fats belong to a group of natural substances called lipids. Lipids are attacked by lipases (lipolytic) enzymes which split the molecule into glycerol and free fatty acids.

Unlike animals, plants can produce all their fatty acid needs by the fixation of atmospheric carbon dioxide using the energy of sunlight (photosynthesis) to keep the chemical reaction mobile. In seed fats the major fatty acid is linoleic, a highly unsaturated fatty acid. Animals produce their quota of unsaturated fatty acid mainly as oleic by removing two hydrogen atoms direct from ingested fat which contains stearic acid.

The complete function of unsaturated acids in animal systems is not known. However, it is certain that lack of these acids in the young animal leads to retarded growth, to water imbalance and to dietary disturbance.

Fish is rich in fatty acid content, the range extending up to C_{22}, and a major proportion are highly unsaturated, some having as many as five double bonds. Aquatic mammals, including the whale, show the same features. In both cases, little is known about their origin or purpose.

Investigations on cooking oils which were subjected to overheating showed that some compounds produced by their decomposition had structures which suggested that they might have toxic properties. On no account should the smoke point of the oil be exceeded. Careful design of apparatus and good management of commercial operations of frying will avoid toxicity and ensure good quality. Packaged dried foods that contain used frying oils on the surface were more sensitive to light than unheated oils. Light induces oxidation, which promotes the formation of rancid flavours.

Organoleptic tests for fat rancidity are often more significant than chemical or physical tests, especially in the earlier stages. A number of standards have been used for the detection of rancidity, these are based on taste and smell.

Rancidity can be caused by: (*a*) enzymes inherent in the tissues; (*b*) odour absorption; (*c*) activity of microorganisms; and (*d*) atmospheric oxidation. The last cause is the most significant due to the unsaturated nature of the triglycerides involved. Hydrolytic rancidity involves the production of free fatty acids with high flavour potency. Butyric and capric acids can be detected by smell and taste in concentrations of less than 80 p.p.m. These free fatty acids may also act as catalysts for oxidative changes. Lipolytic enzymes will be destroyed by heat, therefore this type of rancidity only occurs in products not heated to a temperature high enough for enzyme destruction.

Pure fat will not support the growth of micro-organisms. However, many foods of high fat content also contain moisture and non-fatty tissue. In these, rancid products may develop.

The rate of oxidative rancidity or autoxidation of fats and oils depends upon several factors. Some of the most significant include: (*a*) metals and light; (*b*) surface area of food exposed to the air; and (*c*) available air, temperature and humidity. Copper is twenty times more active than iron. Light activity depends upon the wavelength. Ultra-violet light produces rapid rancidity. In the visible spectrum, red has the least effect and blue the most marked.

Fat shortening ability is measured by the area of flour particles covered by the fat. The greatest shortening power will be possessed by those fats covering the greatest surface area of a chosen flour. Factors such as: (*a*) fat temperature; (*b*) fat concentration; (*c*) nature and concentration of other ingredients; (*d*) extent of the manipulative technique.

Theoretically, physical factors like adhesion, cohesion, surface tension and the molecular attraction between water and organic liquids present in the mixing also have a part to play. When a small amount of oil is placed upon a large area of water it will spread to cover a definite area. This area is governed by the attraction of the polar groups of the fat for water. The oil would be soluble in the water only if the whole molecule were attracted by the water. More energy is required to separate the oil from the water than when polar groups are absent.

A relationship appears to exist between the breaking strengths of pastries when measured by Instron. With increased shortening power the proportion of liquid glycerides present rises and the ratio of liquid to solid glycerides can be used as an index to the quality of commercial shortenings. However, it must be stated that a lot of work remains to be done before one can appreciate all the problems posed when using fats in baking processes.

All the nutrient tests performed on food have a bearing upon quality or legislative procedure. The latter field is dealt with extensively in food legislation manuals. An extract on the meat minima or fat in sausage and other meat products (1967) will serve to illustrate the controls laid down for the protection of the customer:

Product	Minimum meat content	Minimum lean/fat ratio
Sausage, sausage meat, polony or Hogs pudding, other than pork	50	50/50
Pork sausage, pork sausage meat	65	50/50
In a beef sausage and beef sausage meat not less than half of the 50 per cent meat content must be beef. In a pork sausage and pork sausage meat not less than 80 per cent of the 65 per cent meat content must be pork		
Pressed meat	60	60/40
Meat with sauce containing more than 15 per cent onion, mushroom, asparagus, stuffing, dumpling or pasta	40	60/40
Meat with sauce, curried meat with rice, or meat curry with rice.	15	60/40

Note: Definition of meat
For the purpose of food legislation, meat means the flesh, including fat, shin, rind, gristle and sinew in amounts naturally associated with the flesh used.

PROTEIN

It has been known for a long time that protein molecules are made up from the condensation of large numbers of amino acids. In this process, water is split off from the amino group of one acid and the carboxyl group of another. If only two amino acids were involved the result would be a *dipeptide*; the important feature is the peptide linkage indicated by broken lines in Fig. 11.

Note:
By splitting water off as shown the formula for the complex peptides is obtained by a "shunting process"

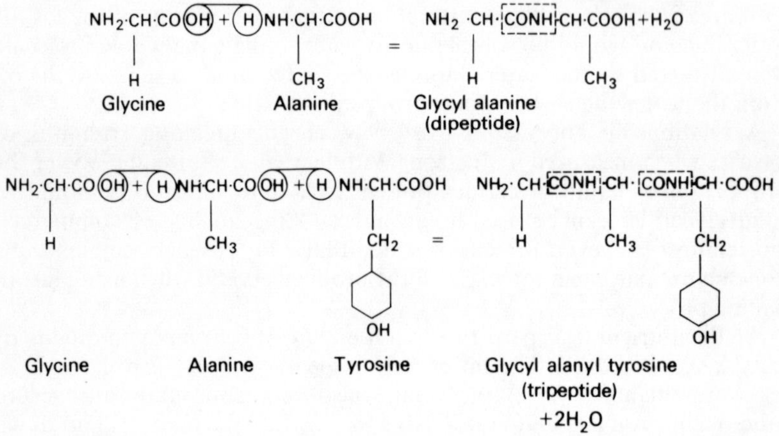

Fig. 11.   Formation of peptides.

When three amino acids are involved a *tripeptide* results and two peptide linkages are present. A natural protein may contain twenty different amino acids which are combined in various numbers to produce a *polypeptide* chain.

All amino acids have the basic formula:

$$NH_2.CH.COOH$$
$$|$$
$$R$$

R can be hydrogen or an alkyl group in the simplest members. In the more complicated members, R will be of aromatic origin as in tyrosine.

Reference has been made to the importance of a diet containing the essential amino acids in the correct proportions. Gelatin and maize are good examples of unbalanced amino acid constitution. Banana ripening shows an increase in such free amino acids as arginine, serine, valine and leucine and a decrease in aspartic and glutamic acids.

Protein denaturation may result physically from the effects of heat, surface forces, freezing, pressure, sound waves and irradiation. The preparation of dried food concentrates demands care if the product is to be easily reconstituted. This is particularly important in the production of milk and mashed potato.

Considerable work has been done on protein denaturation in an attempt to produce food products with optimum protein content and minimum denaturation after processing has been completed.

Denaturation may be thought of as a change in the natural protein whereby it becomes insoluble in solvents where it was previously soluble. The change in solubility of egg white in water heated to 65°C is a noticeable example. The internal structure of protein can be likened to numerous coils or folds of peptide chains which upon denaturation open up to give extended molecules with numerous linkages. Proteins have a complex chemical and physical structure which renders them susceptible to a wide variety of changes, any of which can produce a decrease in solubility.

Heat, freezing, pressure and irradiation are some of the physical means available to produce denaturation. Chemical compounds such as urea and acetamide, organic solvents and ionisable inorganic salts will produce various degrees of denaturation. There are varying degrees of denaturation between the native and completely denatured protein which depend upon the protein in question and the composition of the denaturing agent.

When cooking foods such as meat, fish and eggs, denaturation is easily accomplished. Denaturation in angel cake, bread and other products is an aid to obtaining a rigid framework or texture. Temperature controls the denaturation rate. In some proteins, the heat coagulation at the isoelectric point may increase 500–600 times faster when the temperature is raised 10°C. Coagulation is very rapid at high temperatures; egg white coagulation is a good example. Factors such as the amount of water, salts present and their concentration, presence of sugar and pH affect the coagulation temperature. A dried protein can be raised to as high a temperature as 120°C without losing its solubility. Work on gluten indicates that at low moisture content the rate of denaturation at 80°C– 90°C was negligible, but rose rapidly to maxima at 35–40 per cent moisture.

Custards show that pH, salts and sugar influence the heat coagulation temperature of the proteins present. Sugar elevates the temperature for coagulation of heat protein. Natural proteins have a pH range in which they are stable denaturation may occur if this range is not maintained. High pH is characteristic of deteriorated egg whites. The whites of such eggs may reach pH 9.0–9.5. A higher temperature is required to coagulate this white than the fresh egg white. During surface denaturation, the protein molecule becomes more hydrophobic. Examples of surface denaturation are the insoluble portion of beaten egg whites and the foam on milk. A high concentration of sugar in egg white will prevent surface denaturation. Egg white to which sugar is added before beating is started forms a foam less readily. With high speed beating it is easy to over beat and thus over coagulate the egg white. The egg white is not extensible in the oven and collapses rapidly. However, if sugar is added to the egg white a few seconds after beating begins, surface denaturation is delayed and a better product is obtained.

Amino acids are soluble in water but vary considerably in their solubility. Their solubility also increases as the solution is made basic or acidic; most members are insoluble in organic solvents. These properties are related to the dual basic–acidic or amphoteric character of each amino acid. Each amino acid has a basic amino group and an acidic carboxyl group, and (like common salt) the molecule can ionise.

$$NH_3^+ - \overset{\overset{\displaystyle R}{|}}{\underset{\underset{\displaystyle H}{|}}{C}} - C \overset{\displaystyle O}{\underset{\displaystyle O^-}{\diagdown}}$$

However, the molecule remains intact and electrically neutral and will migrate only when an electrical charge is applied and if the positive or negative charge is made dominant. Common salt has free Na^+ and Cl^- ions which will migrate independently to the appropriate electrodes when a force field is functioning. When the pH of a protein solution alters, either to the basic or acidic side of neutrality, it can migrate under the influence of an electric field.

$$\overset{\overset{\displaystyle R}{|}}{NH_3^+CHCOOH} \underset{+H^+}{\overset{acid}{\longleftarrow}} \overset{\overset{\displaystyle R}{|}}{NH_3^+CHCOO^-} \underset{-H^+}{\overset{alkali}{\longrightarrow}} \overset{\overset{\displaystyle R}{|}}{NH_2CHCOO^-}$$

The pH at electrical neutrality is known as the *isoelectric point*. Above the isoelectric point the protein particle is negatively charged and acts as an anion. Below the isoelectric point it is positively charged and functions as a cation. The isoelectric points of most proteins lies between 4.5 and 7, that of gelatin, for example, being 4.7.

At the isoelectric point gelatin is least soluble, least ionised, and swells the least. In order to prevent cloudiness or turbidity of gelatin in processed tongues, the pH should be adjusted accordingly.

Gliadin is a protein found in the gluten produced when dough is formed. In water it swells to a sticky mass, being least soluble at its isoelectric point, about pH 6.5. With increasing acidity it becomes more soluble, reaching a maximum with a pH range of 2.0–3.0. This accounts in some part for the different rheological properties of dough when treated with varying amounts of ascorbic acid.

Analytical purity, turbidity and surface-tension lowering are some of the properties of proteins which are at a maximum at the isoelectric point. Those at a minimum include viscosity, optical rotation, stability in solution, ash content and osmotic pressure.

MINERAL MATTER

Mineral elements in food and food products are found in organic and inorganic combinations. Malic, oxalic and acetic acids produce salts when combined with sodium, potassium and calcium. At a much higher concentration are inorganic salts, such as the carbonates, chlorides, nitrates and phosphates of the same metals. In addition, magnesium and iron may be part of a complex organic molecule, as in such cases as chlorophyll and haemoglobin.

Calcium is required for a number of vital processes, for example, the coagulation of the blood and the maintenance of the heart-beat. In order to maintain nitrogenous equilibrium, the body must take in at least as much nitrogen as it gives out, as it must of calcium. The average daily calcium loss of an average man living on a diet very poor in the element is 0.45 g, so that unless his food contains at least that amount a 'negative calcium balance' will result.

Vitamin D is essential if calcium is to be absorbed from food. Lack of the vitamin causes the body to extract calcium from the bones, with disastrous results if the drain is long continued. Milk is an excellent source of calcium. An intake of one litre per day will supply about a gramme of calcium. Green vegetables are also fairly well endowed with the mineral.

A cow producing between $0 \cdot 018$–$0 \cdot 020$ m^3 of milk/day can lose between 20–25 g of calcium daily. On good pasture this can be recovered, but on poor pasture, or on defective indoor rations, her diet will be deficient in calcium. The mineral is now transferred from the bones and 'milk fever' or tetany will result, in which disease the blood calcium may fall to one-third of its normal level. Lambing sickness in ewes is identical with 'milk fever'. Pigs appear to be peculiarly susceptible because the requirements of growing pigs for lime salts are very large. The demand for a laying hen that produces 200 eggs in a year is about 0.5 kg.

Requirements of phosphorus, calcium, and Vitamin D are closely interrelated. Brain tissue and fish muscle are rich in phosphorus. This has prompted the fish industry to equate brain power with phosphorus intake. However, the ordinary diet of western civilisation contains a sufficiency of phosphorus, and it is not for lack of that element that our brains are what they are.

Farm animals are in a less secure position. In areas in South and East Africa, and in the United States, the grass does not contain enough phosphorus for sheep and cattle. The result is stunted growth, poor coat, weak limbs and stillborn offspring. Cattle deprived of phosphorus on rough land may devour the bones and carcasses of wild animals and as a result contract 'bovine botulism'. This is caused by a virulent bacillus similar to that which produces botulism in man. It is easily cured by adding phosphorus to the feed.

The integration of copper with iron to prevent anaemia is of interest. Both copper and iron are involved in haemoglobin formation and copper is also essential in the respiratory pigment of crustaceans. Most iron preparations used to cure anaemia contain traces of copper. Pigs have been known to suffer from iron deficiency. In areas where the soil is volcanic in origin iron is lacking and cattle develop 'bush sickness'. This is fundamentally anaemia due to lack of iron.

Goitre, or swelling of the thyroid gland, is due to lack of iodine which is needed to produce thyroxin, a hormone substance. Sea-salt is rich in iodine. Animals living near the sea receive sufficient iodine when salt spray is blown inland to mix with drinking water and to impregnate the soil and the crops it bears.

Apart from the inorganic elements discussed, others such as chlorine and sulphur will be needed. Meat and most vegetable foods contain a relative abundance of potassium and magnesium. Modern farming methods have evolved standard feeds which can supplement any natural deficiency in animals.

Convenience foods for the domestic larder can be produced which have had the correct mineral additives monitored in the manufacturing process. Modern technology has thus eliminated mineral deficiency in the human diet.

VITAMINS

Prior to the First World War work was in progress on a group of chemical substances known as 'vitamines' which were recognised as essential in small quantities to prevent illness or disability. In 1920, Professor Drummond proposed that the terminal 'e' be omitted, because it had been found that none of the vitamins then known really belong to the 'amine' group of chemical substances. Drummond's suggestion was adopted and the word passed into current usage in its present form.

Chemical analysis has to be very accurate if vitamins are to be detected. If they are not present in animal feed their absence soon manifests itself. When three groups of heifers were fed on different cereals, maize, wheat and oats, each group having the standard amount of protein, carbohydrate and fat, the results were as follows. The corn-fed heifers carried their young to full term and produced healthy and vigorous offspring. In the oats-fed group, few of the calves, which were of almost full weight at birth, survived the first few months of life. All the young of the wheat-fed heifers were either stillborn or died within a few hours of birth. The different effects are related to the different Vitamin A content of the cereals consumed. An adequate supply of the vitamin enables an organ to resist infection and restores the growth of young animals.

Beta carotene $C_{40}H_{56}$ which is found in many plants, such as carrots, spinach, tomatoes, and especially in fish oils, is the precursor of Vitamin A. An analysis of Vitamin A shows it to contain a primary alcohol group, its molecular formula being $C_{20}H_{30}O$, which is almost half the original carotene molecule. Beasts have the necessary biochemical agents to produce their own Vitamin A if the feed is correctly adjusted for basic carotene.

Most vitamins are complex in structure but Vitamin C has a rather simple structure. It may be regarded as a carbohydrate derivative. It is distributed widely in animals and plants. Natural sources include Hungarian pepper and the adrenal cortex. It has the structure:

$$
\begin{array}{l}
\text{CO} \\
| \\
\text{C—OH} \\
\| \quad\quad \text{O} \\
\text{C—OH} \\
| \\
\text{HC} \\
| \\
\text{HO—C—H} \\
| \\
\text{CH}_2\text{OH}
\end{array}
$$

The presence of the primary alcohol group shows it to be prone to oxidation. It is unstable to heat. Unlike Vitamin B_2 it is not found in meat and needs to be added artificially if it is suspected that its absence in feeding stocks may cause trouble.

Most of the fifteen or more vitamins which have been isolated appear to act as catalysts or enzymes in essential chemical body changes.

ESTIMATIONS

A large number of routine qualitative and quantitative tests can be devised for nutrient detection. In a qualitative survey the food sample is mixed with water and the following tests applied.

1. Protein—add about 1 cm³ of Millon's reagent carefully (it is a toxic chemical). A white precipitate may be produced. If not, boil for 30 seconds, when a brick-red coloration will indicate protein. This test depends upon the presence of traces of the amino acid tyrosine in the sample. Gelatine gives only a faint reaction as it lacks the acid, but may contain traces of it from other protein substances.

2. Fat—if the fat is visible it will impart a grease spot when rubbed on filter paper. Invisible fat is more difficult and will require solvent extraction. All fat will dissolve in hot caustic liquor to produce soap and glycerol.

3. Carbohydrate—a black stain with iodine proves the presence of starch. A reducing sugar is proved by warming the sample with mixed

Fehlings solution and obtaining an orange or red precipitate. Cupric sulphate from the Fehlings is reduced to red cuprous oxide, the different oxide colours being a result of particle size differences.

A non-reducing sugar such as sucrose needs to be boiled for about 60 seconds with dilute hydrochloric acid. The acid will hydrolyse the sucrose to the reducing sugars glucose and fructose (invert sugars). When the solution is made alkaline and the Fehlings test applied a positive result is obtained.

4. Minerals—these usually occur as improving agents. Examples include bromate, persulphate and phosphate. A sample of the food in water is transferred as a paste to a Pekar plate. When a few millilitres of benzidine are poured on to it, deep blue specks appear if persulphate is present. If the added solution is potassium iodide in the presence of hydrochloric acid, the appearance of black spots will prove a bromate. This test shows bromate to be a powerful oxidising agent, the iodide ion being oxidised to black iodine.

Phosphate is best detected on an aqueous extract of the food sample. When the sample is acidified with nitric acid and warmed with a few millilitres of ammonium molybdate solution a yellow coloration or fine yellow precipitate proves the presence of phosphate. An alternative reagent for the persulphate test is orto tolidine which will produce blue-green specks if persulphate is present.

5. Vitamins—the most important is Vitamin C. To detect it pour Tauber reagent over the wetted Pekar plate and leave for 1–2 minutes. The presence of ascorbic acid is indicated by the presence of bright blue spots or flecks. It is important to examine the plate after the specified time. Reduced iron may interfere with the reaction.

The Pekar test has become established in flour mills. Flour is placed on a piece of glass, pressed down to a smooth surface trimmed neatly and immersed in water. In this way, the colour of several flours can be compared side by side and the test made comparative. The test may also be used to detect bran particles. Catechol reagent is poured over the surface of a prepared Pekar slide. The bran particles contributing to colour grade are immediately stained.

Molisch's general test for soluble carbohydrate (sugars) involves the addition of alcoholic alpha naphthol to an acidified solution of suspected carbohydrate. After a few minutes a violet colour is produced which can be discharged with alkali. Individual sugars will react with phenyl-hydrazine to give osazones of characteristic appearance and melting point.

The Biuret test and the Xanthoproteic test are simple qualitative tests for protein. In the former test, proteins give a purple colour when they are heated with strong alkali and weak copper sulphate solution. The Xanthoproteic test involves heating the protein with concentrated nitric acid, when a yellow stain is produced which goes orange with alkali addition.

Certain physical tests are frequently used to evaluate the quality of a fat for a particular use. In the meat industry fat colour is determined in a colorimeter, using red and yellow glass of known standardised colour characteristics. Melting point, defined as the temperature at which the fat becomes clear and liquid, is obtained by the capillary tube method. Specific gravity and refractive index are two other useful physical criteria.

A useful quantitative test of a carbohydrate nature is the maltose figure of flour. Flours vary in their diastatic activity and hence in their gas-producing power. Flour of adequate activity is required in bread making to produce a good sized, well aerated loaf of good colour. This activity depends mainly on the amount of diastase present and the ability of the starch grains to be attacked. Activity produces maltose which is broken down to provide gas. In the Blish and Sandstedt method flour is incubated with a buffer solution at 30°C and the amount of sugar is determined using alkaline ferricyanide solution. The maltose figure produced is expressed as the number of milligrammes of maltose produced by 10 g of flour. A result of 230 indicates a satisfactory gassing power.

The quantitative amount of reducing sugar in a food sample, for example fondant, may be found by the Lane and Eynon titration method. This involves a titration to give a measure of the amount of reducing sugar in a colourless solution. The standard is a fixed amount of Fehlings solution whilst the sugar solution of unknown concentration is placed in the burette. Heat is applied to the Fehlings and the sugar solution is added until the reactants are colourless and a red precipitate of cuprous oxide appears. The end point is seen more clearly in the presence of methylene blue which is added just before the end point; 10.0 cm³ of mixed Fehlings≡0.05 g dextrose. Typical values recorded indicate 45 per cent invert sugar in golden syrup and 6 per cent reducing sugar as invert in fondant.

Kjeldahl's method for protein estimation in food samples is well established in most quality control laboratories. The experiment is not difficult but it requires experienced hands for safety and reliability. In order to convert the figures obtained from percentage nitrogen into percentage protein, different factors are necessary.

per cent N × 5.7 = percentage protein for wheat flour,
per cent N × 6.38 = percentage protein for milk and milk products,
per cent N × 6.25 = percentage protein for most meat and meat products.

There are three important stages to consider: digestion, distillation and titration. Digestion is a wet, high temperature oxidation process which involves heating the sample in a Kjeldahl flask with concentrated sulphuric acid. During this stage the nitrogen in the protein is converted into ammonium sulphate and other organic compounds are oxidised away. Distillation is carried out after *carefully* making the distilled mixture alkaline. The distillate of ammonia and water is collected directly in a known amount of standard acid used in excess. The final titration is a back titration with decinormal caustic soda using methyl red as an indicator. This measures the amount of unused acid.

$$1.0 \text{ cm}^3 \text{ of } \frac{N}{10} \text{ acid} \equiv 0.0014 \text{ g N}$$

$$\therefore \text{ percentage protein} = \frac{\text{titre} \times 0.0014 \text{ g N} \times 5.7 \times 100}{\text{sample weight}}$$

Digestion   **Distillation**

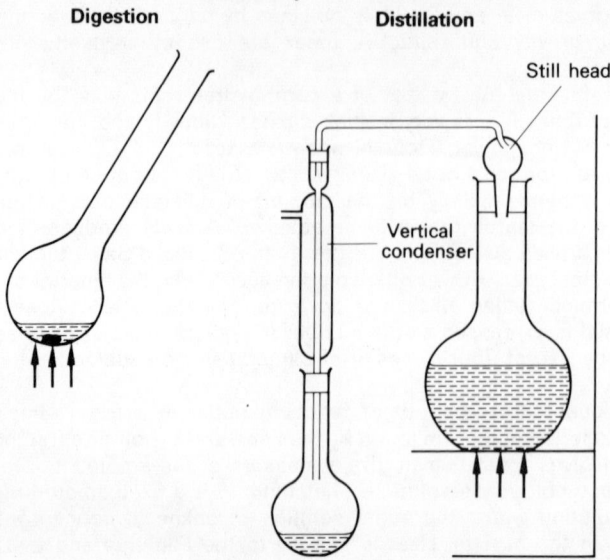

Fig. 12.   Kjeldahl protein determination.

The calculation would be applicable for wheat flour. Figure 12 shows the apparatus involved.

A quantitative analysis of the fat content of food can be determined using the Soxhlet apparatus. The Soxhlet extracting unit contains about 5 g of the sample in a dry state, wrapped in filter paper and held in place with aluminium wire. Above the unit is a vertical condenser and below it a distillation flask containing a permitted solvent, usually petroleum ether. The ether is distilled, condensed and allowed to penetrate the food continually. Ether and fat leave the extraction unit by a siphoning action. The solvent is recovered by distillation and the weight of the fat, left as a translucent deposit in the distillation flask, is recorded.

$$\therefore \text{ percentage fat} = \frac{\text{fat weight} \times 100}{\text{sample weight}}.$$

The Reichert value distinguishes between true butter and imitation products. Butter fat contains a fair amount of butyric acid glycerides and the solubility and volatility of this acid is used in a characteristics value determination. The value is defined as the number of millilitres of decinormal alkali solution required to neutralise the volatile soluble fatty acids distilled from 5 g of fat. In the test butter fat is first separated from butter. The pure fat is saponified, and then distilled after making the mixture acidic. The distillate is then titrated against standard alkali. The Reichert value of butter varies between 24 and 32.

The Rose–Gottlieb process is similar to the Soxhlet technique and is particularly suitable for milk and some milk products, for example cream. The food is treated with ammonia solution and alcohol and the separated fat is extracted with an ether-petroleum mixture. Alcohol is used to precipitate any protein which dissolves in the ammonia allowing the fat to separate. Trade cream for whipping must contain approximately 40 per cent fat, and double cream not less than 48 per cent. One would expect cream substitutes to contain at least 25 per cent fat and fat oil.

In routine work the Gerber method for milk is used and is very convenient. Special cream butyrometers are available and the test usually involves 1.1 g of cream.

$$\text{Percentage fat} = (10 \times \text{butyrometer reading}) - 1.$$

The stability of oils and fats can be measured either by (a) peroxide value of oils and fats or (b) acid value of oils and fats. Both are volumetric methods. By allowing a known weight of oil or fat to react with potassium iodide and titrating the liberated iodine against standard sodium thiosulphate solution the peroxide value can be determined.

The value is defined as the number of millilitres of $\frac{N}{500}$ sodium thiosulphate which is equivalent to 1.0 g of sample. For good quality a peroxide value of not more than 10 is desirable.

Hydrolysis of a fat, due to enzymic activity, produces free fatty acids which can be estimated by titration against standard alkali. The acid value figure is defined as the number of mg of potassium hydroxide required to neutralise 1 g of sample, the acid used in the expression is oleic. An acid value not greater than 2.0 is desirable.

When determining the ash content of a food it is usual to allow a known portion to burn, and to weigh the residue. The organic portion disappears and the inorganic remains, usually as a white ash.

Ash content figures are used to measure quality, and to detect adulterants and foreign matter. Ground pepper quality is related to ash, black types giving a high result. Plain flour figures are no longer related to quality due to compulsory additions of mineral matter. A list of expected figures is as follows:

High class patent flour	0.3 per cent
White pepper	1.0 per cent
Cow's milk	0.8 per cent
Beef	1.0 per cent
Sausage	2.5 per cent

Quantitative vitamin analysis is complex and beyond the scope of this book. A number of the methods used involve photographic spectrophotometry and direct vision spectrophotometry. In a number of large industrial concerns vitamin analysis involving photoelectric spectrophotometers is well established.

USES

Carbohydrates are energy foods which in the form of sugars can be regarded as flavouring agents. Commercial glucose, or liquid glucose prepared by hydrolysing maize starch, is widely used in the manufacture of boiled sweets and fondant. Approximately 2 lb sucrose/week is consumed either as pure sugar or in finished goods for each person in the United Kingdom. Invert sugar is found in jams and sweets and in the sweetening of fruit drinks. Chocolate may contain up to 50 per cent sugar and the total carbohydrate value of plain chocolate is in the region of 60 per cent.

In animals glycogen acts as a store of carbohydrate and it is sometimes referred to as animal starch. Cellulose has no nutritional value to man but can act as a roughage material. Baked beans contain a portion of hemicellulose which can act as a stimulant to the alimentary canal.

Dextrins give colour and texture differences to food. Most children appreciate toast rather than plain sliced bread, while crusty bread is also an enjoyable dextrin product. Pectins, as mixtures of polysaccharides, are essential to the formation of gels in jam manufacture. Bread, flour, rice and oatmeal all have carbohydrate contents of more than 50 per cent.

Proteins represent the most complex units known. Lactoglobulin in milk has a molecular weight of approximately 42 000 which dwarfs the 342 for sucrose. Muscle, skin tissue and enzymes are of a proteinaceous nature. Because of the infinite number of chemical combinations the elasticity of proteins is considerable. Unlike carbohydrates the world's supply of protein is under considerable pressure. One reason is that animals must consume between 1.5 and 2.0 kg of protein in order to produce about 0.5 kg of protein for human consumption.

Recently the potentialities of protein from vegetable, rather than animal, sources have become public interest. In developing countries it is estimated that cereals provide up to 70 per cent of the protein intake. Efforts are being made to increase the protein quality and quantity by genetic selection. Success has been achieved with certain maize strains to produce a high lysine variety. Soya protein provides ten times more lysine than wheat; in addition the oil fraction has useful soluble vitamins, antioxidants and lecithin.

The introduction of 'Kesp', an edible spun protein fibre from soya, is available in three basic forms: mixed chunks, pieces or minced, in flavours resembling chicken and beef.

Tests carried out by the author, in which the public were asked to select pure beef from pure soya protein and mixtures of the two in a baked product, indicated that only approximately 30 per cent selected the pure beef as true meat. Providing the preparation of soya protein is expertly carried out there is little doubt that an acceptable meat analogue can be prepared. This is not to say that the flavour and texture equal that of lamb or beef, but 'Kesp' has much to offer as a source of good protein. Figure 13 illustrates the nutritional value of spun protein against stewing steak.

In Sweden the world's first plan for manufacturing high protein con-

Nutrients in soya beans for 100 g

Protein 40 g
Fat 20 g
Carbohydrate 24.5 g
Balance Ca, P_4, Na, Cl, K, Fe, Mg. Vitamins A, B_1, B_6, C, D and E
Calorific value 1965 kcal per 450 g

Nutrient analysis of 'Kesp' and stewing steak (1 oz portions)

	'Kesp'	Stewing steak
k. cal	80	61.5
protein (g)	6.2	4.8
fat (g)	5.9	4.5
carbohydrate (g)	0.03	0
calcium (mg)	3.4	3
corn (mg)	1.2	1.1
k. cal supplied by protein (%)	32	32

Fig. 13. Nutritional value of soya protein.

centrate from rapeseed has recently been completed. The process is completely new and the pilot plant is producing protein for animal consumption. If all goes well the results of the project will permit the exploitation of a new protein source for human consumption.

Fats, in addition to being a concentrated slow burning energy source, have the ability to provide flavour to food, even though they themselves have no flavour in the true meaning of the word. Chipped potatoes, fish fingers and fried bread illustrate this point well. Superglycerinated or high ratio fats are ideal emulsifiers for use in cake making. Fats have different creaming powers; in general the more saturated the fat the less its ability to incorporate air although other factors are involved as well. The use of fat as a 'shortening agent' in biscuits and shortbread is well established.

Minerals are present in small amounts as the ash content of food; and, as previously mentioned, their function in the body is widespread, from bone structure and cell fluids to conjugated proteins. Likewise the vitamins, which are present in even smaller amounts, are needed to combat the unpleasant symptoms of nervous disability, scurvy, rickets, etc. Fruit and vegetables are one of our main source of vitamin intake. In the case of Vitamin C they provide almost 90 per cent of our daily quota, whilst over 40 per cent of our Vitamin D source is pure fat.

EXERCISE 5

1. Which vitamins are incorporated in margarine? Explain their functions and examine the vitamin aspects of:
   (a) cheese;

    (*b*) meat; and

    (*c*) fresh fruit.

Discuss briefly any essential conditions that might be required to preserve vitamins.

2. What basic differences exist between fats and carbohydrate? Illustrate your answer by reference to a comparison of chemical, calorific and nutritional aspects of both substances.

3. Explain the term peptide linkage. To what extent does the isoelectric point of a protein account for its physical properties?

4. What particular minerals would you consider of value in meat and meat products? Discuss the role they play in a healthy beast.

5. Distinguish between water soluble and fat soluble vitamins. Indicate their importance in the production of healthy breeding animals.

6. Describe the qualitative tests necessary to discover the nutrients in:

    (*a*) ox liver;

    (*b*) plant bread; and

    (*c*) fruit tart.

From the analysis what conclusions would you draw as to their inability to produce a balanced diet?

7. How would Kjeldahl's protein estimation be of value in comparing joints of meat from a carcass?

    2.5 g of sausage was analysed by the Kjeldahl method. The released ammonia was passed into 25 cm³ of $\frac{N}{5}$ H_2SO_4. The unused acid needed 10 cm³ of $\frac{N}{10}$ NaOH for its neutralisation. Calculate the percentage of protein in the sample. In your opinion is the protein content of acceptable quality?

8. What are the scientific principles behind the Lane and Eynon estimation of reducing sugars in confectionery products?

    5 g of fondant were dissolved in 100 cm³ of distilled water. Three titrations, 22.5, 23.0 and 22.9 cm³, were recorded in discharging 10.0 cm³ samples of mixed Fehlings solution. What percentage of invert sugars estimated as glucose are present and how would they affect the physical properties of the fondant?

6 Food colours and flavours

INTRODUCTION

Man lives with colour, and colour is part of his daily experience. Indeed a life without colour is difficult to imagine. It is true to say that man eats with his eyes as well as with his mouth. Natural, living colours stimulate the appetite and digestion and increase the pleasure of eating. The physical nature of colour is dealt with in Chapter 11.

It is no accident that for centuries saffron, extracts of berries and other naturally occurring food colours have been used to advantage in food. In New York white hens' eggs are in demand whilst in Boston brown eggs reign supreme. Whilst the egg shell colour is significant in some parts of the world, in others fat colour is important.

Carotene from grass becomes concentrated in fat, and in the summer while milk has its full quota, butter produced has associated with it the acceptable colour of yellow favoured by the British housewife. When animals are fed on indoor fodder in the winter months the carotene level falls and the resulting butter is very pale. In extreme cases it would resemble lard and sales would suffer considerably. The judicious application of permitted colour rectifies this.

In the last twenty years the total synthesis of B-carotene has made its incorporation into a wide variety of products possible. The carotenoids are natural pigments which are responsible for the colours of butter, eggs, carrots, green vegetables, tomatoes, lobsters and apples. Scientific theory and practice have shown that in this instance Nature has been successfully imitated. For this reason the carotenoids are completely harmless, and as a result are widely used for the safe reliable colouring of foods. As little as 3–5 g of B-carotene will give 1 tonne of margarine an appetising golden colour as well as providing a precursor for Vitamin A production.

There are basically three classes of food colours available to industrial application: (*a*) natural; (*b*) permitted synthetic; (*c*) inorganic or mineral pigments.

Natural colourants include cocoa, annatto, caramel, saffron, cochineal and carbon black. These may be obtained from animals and plants.

Cochineal, which colours food red, is obtained from the bodies of dried insects whilst annatto, which colours cheese and turmeric mustard pickles, is obtained from the annatto tree by crushing and extracting the fruit with warm water. Caramel colours certain drinks and flour confectionery products.

The permitted synthetics include the carotenoids and coaltar dyes, the latter having no equivalent in nature at all. For products high in shortening, the carotenoids are very serviceable and give good uniformity.

B-carotene can be used for butter sauces in vegetables, sold either in frozen form or canned products, where it is important that the vegetable should not absorb any of the colour. This is an advantage over the coaltar or azo dyes which will eventually tend to colour only the vegetable. The azo dyes offer a wide variety of shade which cannot be obtained from natural sources but are subjected to exhaustive tests before they can be released for general use. A synthetic colour cannot be regarded as completely safe but is taken, as it were, on trust. The formation of carcinogenic compounds is always a possibility and agreement on this is by no means universal. For example, the Swiss list of artificial colouring agents permitted for use in foods was reduced from twenty-eight to twelve in 1957 and in the United Kingdom thirty colours were allowed until recently, but the Foods Standards Committee recommended in 1964 that six of them should be withdrawn. This recommendation has been embodied in the 'Colouring Matter in Food Regulations 1966'.

Included in the inorganic colours are the aluminium lakes, iron oxide, titanium dioxide, silver, gold and aluminium. The aluminium lakes are well suited for colouring icings rich in shortening, and in sugar icings to eliminate bleed into white or very light cake. Lakes are also useful where the shade of crust and crumb must be similar. Most inorganic colours are stable in changing conditions but are dull, limited in shade and insoluble in water. Inorganic colours have only a minor use; examples include iron oxide (aniseed balls) and 'silver' cachous used for decorative work on sugar-coated confectionery.

Flavour, like colour, has a considerable influence upon our appreciation of food. Stale eggs or fish may possess the same nutritional value as the fresh products but do not appeal to the appetite. Fat is prone to rancidity and even traces of butyric acid will make stocks of butter unpalatable.

The term flavour has been used in various ways, but is usually applied to the over-all sensations obtained by odour, taste and touch. The feel of some foods is also an important factor in their palatability. Examples of this are the crispness of crackers and certain types of cookies, the smoothness of custards and the velvetiness of cake crumb.

Organoleptic tests are an important aspect of consumer research. Tests are of two types, preference and difference, margarine versus butter being a good example. Some food firms or processors have housewives, living in different parts of the territory in which the product is sold, to test new and old products. The coded products are dispatched to the housewife in unlabelled, plain containers.

The skin and peels of fruit carry the most flavour, pressure being necessary to obtain the maximum yield. Hydraulic presses operating at about 300 kg/cm^2 are used to extract ginger essence from solvent-treated material. Flavour is at a maximum when the skin is at its most brilliant.

A particular flavour is the result of a large number of chemicals combining to produce a sensation which will stimulate our taste buds. As an example of the complex nature of flavour one can examine the aroma of white bread. The volatile constituents in the vapour were investigated using gas chromotography and mass spectrometer techniques. Extraction of the volatiles under vacuum from an aqueous slurry of white bread allowed no less than *fifteen* components to be identified in the vapour and *twelve* in the aqueous extract!

Similarly, the isolation of volatile components from beef broth using infra-red absorption spectrometry and gas-liquid chromotography revealed a large number of chemical components, among them higher alcohols and benzaldehyde. The production of synthetic flavours to compete with the natural ones appears to present problems of a greater complexity than those faced by the food colourist. However, continued research assists in unravelling them.

COMPLEX NATURE

Neither colours nor flavours contribute any significant food value to the items containing them and yet both are essential if food is to be of succulent quality. It is not permissible to add colour to any fresh food source such as meat, poultry, fish, fruit or vegetables. White bread, cream, condensed milk or dried milk are also subject to the same restrictions. Colour may be added to the husks of nuts or the skins of oranges, but any edible food portion is protected by law.

The use of an additive which may restore the original colour of fresh meat is prohibited. Colour in meat is due to the presence of two conjugated protein complexes, haemoglobin and myoglobin. Haemoglobin is concerned with blood and myoglobin with meat muscle. Myoglobin may be regarded as the primary pigment; it is purplish-red in colour and, depending upon the pressure of available oxygen, it can either combine with oxygen physically (oxygenation) or be oxidised chemically (oxidation).

The element iron is in the *ferrous* condition in both myoglobin and oxymyoglobin but in the *ferric* condition in the oxidised form metmyoglobin. Minced meat rapidly loses its bright red colour as it undergoes oxidation to brown metmyoglobin. Colour is, therefore, indicative of the age of meat. If a reducing agent such as ascorbic acid is rubbed on to the surface of old meat the colour will be restored to that of fresh meat, owing to the reducing action of the acid.

In the interior of meat the myoglobin is in the reduced state and tends to have a purple colour. Enzymatic oxidation of glucose substrates in meat will produce coenzymes capable of keeping meat looking fresh. After death, the tissues' supply of oxidisable substrates ceases, the reducing power of muscle is lost, and the iron is oxidised to metmyoglobin. Figure 14 illustrates the chemical changes involved.

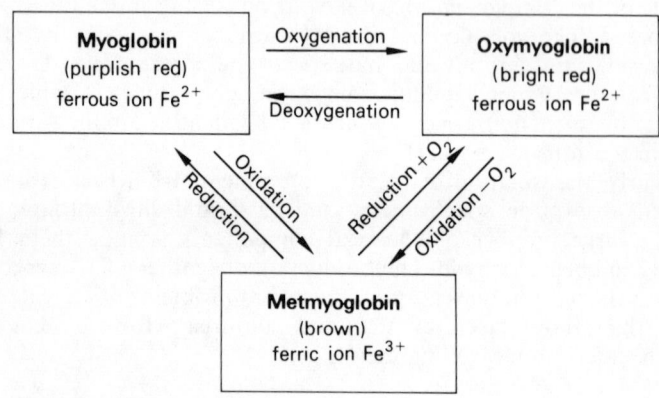

Fig. 14. Main colour mechanism in meat.

Meat may occasionally take on a greenish tinge which could be confused with mould growth. In fact, the pigments involved are due to bacterial action producing sulphomyoglobin and cholemyoglobin. Both pigments are prone to oxidation and protein denaturation to complex molecules called porphyrins. These contain heterocyclic aromatic units involving nitrogen. The meat will assume a yellow or brownish tinge reminiscent of metmyoglobin.

The stable, pink colour of cured meats is due to the combination of nitric oxide from the curing bath forming a covalent complex with myoglobin to produce nitrosomyoglobin which, unlike haemoglobin, is unaffected by temperatures of 100°C. All these colour complexes can be identified by their absorption spectra. For example, metmyoglobin has a large peak in the blue end of the spectrum and a weaker peak in the red. The net result is to produce a visible brown compound.

The manufacturers of food products try to retain as much of the normal colour of the raw product as possible in the finished state. Natural colours are affected by temperatures met in canning, and by enzyme impact, light and pH changes, etc. Some colour loss or change is therefore to be expected. Products which have to be frozen or dehydrated are given a preliminary blanch which will inactivate enzymes that could affect colour and also drive out dissolved oxygen which could produce colour loss. For these reasons artificial colours are preferred in order that colour uniformity may be preserved.

Cherries containing natural colour are not identical, even from the same bunch. The use of approved bleaching agents, followed by permitted colour additive, will produce in the canned product a homogeneous result, and the same treatment may be applied to green peas.

Snack food manufacturers are interested in synthetic colours for use in extruded snacks. Normally, colours are available at the coating stage and not added internally at the time of extrusion. Experiments are being conducted to evaluate suitable colours and conditions for this to be a reality, so bringing commercial marketing advantages.

Colours used in confectionery should have good heat stability, and resistance to sulphur dioxide and the effects of direct light. It is always preferable to add the dye at the end of a cooking process, both to limit the time of exposure to a high temperature and to minimise any sulphur dioxide effect that may have been present initially as a fruit pulp preservative or in cane sugar/glucose mixes. With jellies, the gelatin will contain sulphur dioxide traces; in addition, any added colour must be able to resist a low pH and have a good solubility.

Precipitation by fruit acids may be an advantage in certain instances, as non-bleeding coloration can be obtained. For example, bleed resistance of Erythrosine BS in cherries can be achieved; hence their use in fruit salads.

Lacquered cans are to be preferred to unlacquered ones because of the reducing action set up by the interaction of the tin and iron leading to 'hydrogen swell'. Unlacquered cans can be used for tomatoes and no

Food manufacturing process/food type	Sunset yellow FCF	Tartrazine	Amaranth	Carmoisine	Erythrosine BS	Red 2G	Ponceau 4R	Red 10B	Indigo carmine	Green S	Brown FK	Chocolate brown FB
Sugar confectionary (boiled sweets, fondants)	×	×	×	×		×	×	×				×
Preserves (jams, jellies)	×	×		×		×						
Canned fruits and vegetables		×		×	×	×				×		
Soft drinks and essences	×	×		×		×	×					×
Baked items (cakes and biscuits)	×	×			×	×	×			×		×
Dry blend applications (custard and blancmange powders)	×	×	×				×					×
Meat products (sausages, pork)						×	×					
Fish pastes (salmon)					×							
Cured fish	×	×									×	
Ice cream	×	×		×		×	×					×

Fig. 15. Applicability of colours to various processes.

synthetic colours are applicable in this case. The colours used in soft drinks must remain stable when exposed to sulphur dioxide, acids and light.

Increased interest is being shown in water-insoluble calcium or aluminium lakes of permitted colours as they have the advantage over water-soluble colours in fat-based materials such as biscuit creams and marzipans. They have the added merit that, because they are insoluble, they do not stain the mouth. Methyl violet is not a permitted edible dye but can be used as a marking agent on the skins of citrus fruits.

Figure 15 gives a list of some of the permitted colours in various foods with their proprietary names.

On a qualitative basis, the distinction between a natural, permitted synthetic and non-permitted basic dye involves dyeing a piece of wool in an acid bath of the colour. Coal-tar dyes normally dye animal fibres readily, producing a bright colour. This can be removed by stripping in alkaline conditions. When a natural, organic colour dyes the fibre, it will not, as a rule, impart a bright colour; neither will the colour strip 'clean' from the fibre as a dirty brown to purple colour remains. A non-permitted basic dye such as Rhodamine B will dye from an acid bath but produces a strong fluorescent effect.

Such basic dyes are carcinogenic. Rhodamine B, if used in colouring rock, would produce cancer of the mouth under optimum conditions. Rhodamine B is an exceptional basic dye in being able to dye from acid conditions. Normally, if the wool remains white, either no artificial dye is present or (an unlikely possibility), one of a basic type has been used.

Modern colour identification employs colorimetric, chromatographic and spectrophotometric measurements. The colour strength of caramel may be compared using two 50 cm³ Nessler cylinders. Quantitative dilution of the original sample up to about 100 times is usually necessary. The colour produced can be visually compared against a known standard of centinormal iodine which is equated in terms of a known caramel concentration.

A similar technique for brine curing bath strengths is used. This is based upon the ability of nitrite to produce an unstable diazonium complex which, with a naphthol, gives a pinkish hue to an aqueous extract of known concentration. By matching this against known standards, an accurate estimation is made of nitrite as p.p.m. Colorimetry is discussed at greater detail in Chapter 8.

The spectrophotometer is the most accurate instrument tool for colour measurement. It eliminates human errors in colour matching and is quicker in use than thin layer chromatography. It permits the analyst to measure the amount of light absorbed in all portions of the spectrum by a coloured substance. In the case of food colours, the absorption usually occurs in well-defined bands. The amount of light absorbed is plotted against wavelength and an absorption curve is obtained. Under controlled conditions, the spectral positions and general forms of the curve are characteristic of individual dyes and the height of the band is a direct measure of dye concentration.

Flavour is detected by an accumulation of information from the olfactory organs and the taste buds. Four basic tastes exist, salt, sour, sweet and bitter. The tongue has definite areas associated with the different tastes, salty and sour tastes being related to the sides, sweetness to the front and bitter to the back. There is no absolute quantitative method available for flavour evaluation. Vacuum fractional distillation can separate flavour components from food, and paper or gas chromatography will identify the substances constituting a particular flavour. However, the final appreciation is primarily measured through human sense response to stimuli, and this response will vary with the individual.

Flavouring agents may be natural or synthetic. Originally, most flavouring agents were natural products added to make an insipid food more palatable or to make a deteriorated food product edible. Pepper, ginger, onions and garlic are able to enhance flavour and may also be used incorrectly as a means of counteracting bad flavour caused by rancidity. Today, the trend is towards blended or harmonised flavours to make the food more interesting, it being realised that flavour does most to relieve monotony in the diet.

Natural essences are prepared by macerating natural flavouring materials in appropriate solvents. Roots, seeds and fruits may be treated with various alcoholic compounds, including ethanol and isopropyl alcohol. Essential oils are obtained by expressing the oil from the peel or fruit or by distillation in steam. The latter process produces oil of low quality due to the presence of terpenes. These are hydrocarbon substances which are colourless liquids with a pleasant smell. Limonene $C_{10}H_{16}$ occurs in the essential oils of oranges and lemons. Artificial essences are prepared by blending various organic compounds with a suitable solvent.

One of the great virtues of artificial essences is their ability to compete very effectively with the natural product. Synthetic vanillin is much cheaper than the natural substance, which takes two years to mature, and since the flavour is relatively simple it has found great use in the confectionery business. However, most imitative flavours do not fully agree with the natural product; there are subtle differences. This is hardly surprising in view of the complicated nature of flavour. An example of this is imitative apple flavour which may have upwards of twenty-two different compounds. These can include such diverse components as benzyl formate, butyric acid, aldehyde C_{16}, acetaldehyde, propylene glycol and amyl butyrate.

Fresh meat from young animals has a very slight odour of lactic acid which has resulted from glycogen degradation. Older animals, particularly mature male animals have a stronger odour. The presence of boar taint can be tested by applying a battery-operated soldering iron to the fat part of a carcase. An unpleasant smell of ammoniacal origin is released where the hot tip is in contact with the flesh. Investigation into boar taint has resulted in a practical aid to pig farmers using artificial insemination. The odour is produced by the submaxillary glands amongst the three salivary glands. By 'capturing' the compounds responsible it is possible to stimulate an otherwise reticent sow or gilt. The substance is now in an aerosol can and may be thought of as a 'reproductive catalyst'.

A second 'spin-off' from flavour investigation concerns a synthetic raspberry-flavoured ketone. By acetylating it, a compound has been produced which, when dispersed in fine droplets in the air, prevents the mating of certain insects. Unlike the organo chloride pesticide complexes, it is harmless to animals. It contributes considerably to pest control.

The aroma condensates of cooked meat include ammonia, amines, hydrogen sulphide and short-chain aliphatic acids. Under unfavourable storage conditions, unfrozen meat can develop proteolytic or putrid odours from protein breakdown, sour odours from microbial growth, and rancid odours from fat oxidation. These 'off odours' are described as oily for pork, tallowy for beef and muttony for mutton. Meat stored after cooking can develop a stale odour. It is possible that this is due to myoglobin-catalysed fat oxidation after cooking.

In certain cases, pure chemical compounds can add flavour. Cyclamates have now been banned from soft drinks but saccharin has stood the test of time as a useful alternative sweetener to sugar. Monosodium glutamate is an olfactory catalyst for many meat dishes. As a result of considerable research into the flavours produced during the cooking of meat, the inception of flavour boosters is a practical possibility. These flavours may be produced in eight different types including three varieties of beef and are available in paste or powder form for use in sausages, pies, canned and packet soups, and many other meat products.

The esters form a good basis for fruit flavours and are suitable for application to fondants, high boilings, table jellies, pectin jellies, and blancmanges. Butter, toffee and nut flavours are produced in concentrated form and are economical in use.

The growing demand for coffee whiteners, imitation creams, and cheese-flavoured products has created a demand for suitable dairy flavourings. Enzymatically developed flavours can be produced in relatively high concentration and closely match the aroma of the natural products. The aroma of ripened cream and butter from Japan and the United Kingdom is quite different, due to subtle changes in the profile of certain free fatty acids. These differences can be satisfactorily simulated in the enzymatically developed product. Such products are finding an increasing application in the various cheese-flavoured snacks. It is often a most disconcerting experience for a manufacturer who uses an excellently flavoured cheese powder in the manufacture of a snack, to find that the product has very little distinctive cheese aroma and flavour. These difficulties can be successfully overcome by the use of enzymatically developed cheese flavours, carefully matched for the type of product and cheese used.

Problems facing flavour manufacturers will vary depending upon the individual section of the food industry. In the confectionery field, the flavour constituents should not be affected by heat, light acids or water, and should be able to blend well in a sugar medium.

The solubility rate of the product is an important factor in dosage levels. Toughness or chewiness of a product largely determines the time the sweet is in the mouth and therefore affects the dosage of flavour required. Gelatin jellies will always require a higher flavour dosage than pectin or agar jellies.

Caramels, toffees and chewing gums contain fats and gums that absorb flavours and therefore retard the rate of dissolving. Chewing gum is masticated for an average time of 5 minutes, and therefore the level of flavour required is approximately 10 times that required for a boiled sweet.

The size and specific gravity of the confection also affects the dosage of flavour liberated per mouthful per second. Marshmallows whipped to a specific gravity of 0.4 would require approximately a third more flavour than those whipped to a specific gravity of 0.8, to give the same taste impact.

Particle size of chocolate affects its flavour. Above 25 μm the particle will be detected on the palate and a poor flavour impression received. It is essential that any milk powder used in chocolate production is free from active lipase. Lipase can split the fat into soapy-flavoured fatty acids which, when produced, will not be masked by flavour additions. Careful selection of any flavouring solvent is necessary, particularly with regard to boiling point or possible action on any of the raw materials incorporated in the production of a sweet line.

Finally, it must be emphasised that good liaison between the flavour suppliers and the manufacturers will save time, and can yield mutual advantages.

SYNTHETIC EQUIVALENT

Coal tar provides the raw material from which the intermediates benzene, naphthalene and anthracene can be obtained. Naphthalene and anthracene, when treated with chemical reactants containing essential groups, are converted into permitted food colouring additives. Diazo compounds which contain the azo group can, by coupling techniques, produce stable molecules. As the molecule becomes more complex the colour deepens through the spectrum range. The triphenylmethane dyes $(C_6H_5)_3CH$ have a minor role to play in synthetic equivalents; they have one advantage over coal tar dyes in their ability to offer more resistance to fading in the presence of tin dissolved from tin-lined containers or tin-lined vessels used in the manufacturing process. Triphenylmethane dyes such as Fast Green, Guinea Green and Brilliant Blue are used either individually or in selected combinations.

Carmoisine is an example of the type of structure encountered in azo dyes. The coupling component is a naphthol sulphonic acid in the form of a sodium salt.

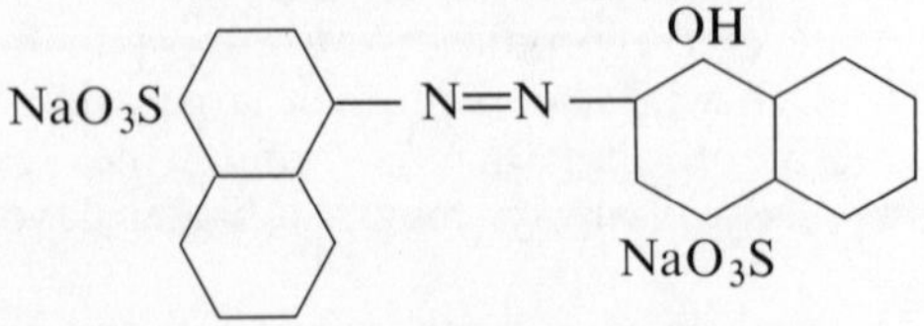

Carmoisine

All the coal-tar dyes in foods at the present time are water soluble. They are all substantive dyes in that they dye animal fibre (wool or silk) without the use of a mordant.

The chemical structure of synthetic flavours is initially less complex, a number being substituted benzene rings. Figure 16 shows that the main flavouring component in a number of food examples is often aliphatic in nature.

Food or food use	Principal flavouring component
Butter	di-acetyl
Orange	acetyl acetate
Raspberry essence	pentyl acetate
Onion	ethyl thiocyanate
Garlic	diallyl disulphide
Banana essence	pentyl butryate
Pear	propyl acetate
Almond oil	benzeldehyde
Apple essence	ethyl acetate
Coconut	gamma-nonalactone
Rum essence	ethyl formate

Fig. 16. Synthetic flavouring agents.

Monosodium glutamate is based upon the non-essential amino acid, glutamic acid. The side-chain R is not unduly complicated.

$$NH_2.CH.COOH$$
$$|$$
$$(CH_2)_2$$
$$|$$
$$COONa$$

Monosodium glutamate
(accentuates meaty flavour)

Other examples include:

$$(CH_3)_2C\!:\!CH.CH_2.CH_2.C(CH_3)\!:\!CH.CHO$$

(Citral: Lemon flavour)

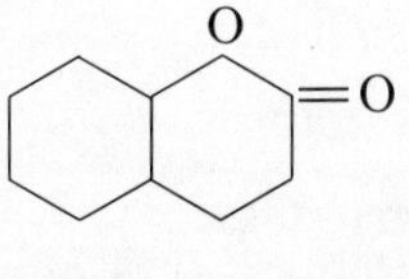

(Coumarin: Bad vanilla) 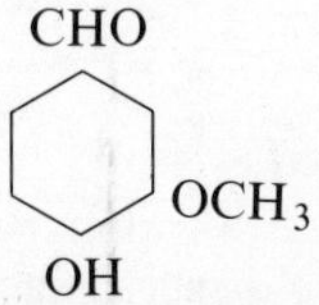   (Vanillin: Vanilla) 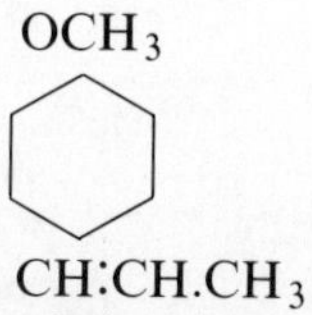   (Anethole: Aniseed)

and $CH_3CO.CO.CH_3$

(Diacetyl: Butter)

IMPORTANCE OF UNSATURATION AND SPECIFIC REACTIVE GROUPS

A compound will be coloured if it contains *chromophore* groups. These are colour-bearing groups. Compounds which contain one double bond do not produce colour, but when several double bonds are present, in conjugation, colour may develop.

Some of the more common chromophore groups are:

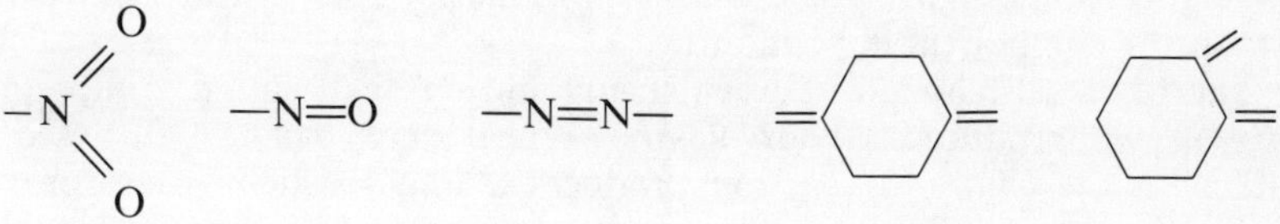

Azo benzene $C_6H_5.N\!:\!N.C_6H_5$ crystallises in brilliant red plates but will not dye a substance. Such compounds are known as *chromogens*. If it is reduced to hydrazobenzene $C_6H_5.NH.NH.C_6H_5$ the colour is destroyed as the azo linkage disappears. The source of colour is related primarily to a deficiency of electrons in the molecule. For example, completely saturated compounds (organic) show no selective absorption throughout the visible and ultra-violet regions.

With few exceptions, all hydrocarbons and aliphatic compounds composed of carbon, hydrogen and oxygen only are colourless. Quinones are distinctly coloured although they contain only carbon, hydrogen and oxygen, the explanation lying in the double bond between carbon and oxygen. Upon reduction they are converted into colourless derivatives.

Advanced food science

Quinone (coloured)   Hydroquinone (colourless)

All substances possess absorption bands, but these bands do not necessarily fall in the visible region of the spectrum. It is only those substances which possess absorption bands in the visible range that appear coloured.

The presence of reactive groups in a molecule can often be associated with its particular flavour. Only general guidance can be given, such as aldehyde, sharp and ketone, pleasant. Carboxyl and sulphonic acid groups will sharpen flavour and the latter group also imparts good solubility. The hydroxyl group has a mellowing influence whilst the amino group exercises 'off-flavour' influences.

Molecular size is also another factor affecting flavour. Acetaldehyde has a pungent odour whilst benzaldehyde is quite pleasant. Aliphatic ketones are pleasant substances, but in coumarin the aromatic ketone grouping exerts the opposite effect. Indeed, if food contains coumarin it will be regarded as adulterated, since in 1953 a commercial laboratory reported that coumarin possessed some toxic properties.

Acetic acid is regarded as a preservative rather than a major flavour factor; it is quite sour. With the introduction of an hydroxyl group into its higher homologue, lactic acid is produced. This is the principal flavouring component in yoghourt.

The Maillard reaction between sugars and terminal amino groups in proteins will produce a savoury flavour. Wheat gluten can be hydrolysed, and with the Maillard reaction produces a liquid which, after spray drying on to cornflour, gives a meaty, solid flavour. Para hydroxyl benzaldehyde can be treated chemically with reactive groups of ketonic origin to produce a raspberry flavour base.

The interaction of basic flavour groups is the prerogative of commercial flavourists who naturally are reluctant to reveal their secrets. A large number of molecules can be produced with good flavour potential but may possess a latent carcinogenic factor. Vanillin illustrates clearly how the reverse is applied. Benzene is highly toxic, but as a trisubstituted derivative of it vanillin has become a highly desirable synthetic flavour.

EXERCISE 6

1. Write a short account of the limitations of natural colours in commercial food production.

2. What chemical reactions underline the formation of azo dyes? Why have they been the most successful synthetic food colours?
3. Discuss the practical steps necessary to identify the colouring component in canned peas.
4. What factors are in favour of the production of a synthetic flavour for food?
5. To what extent may flavour be regarded as a chemical, rather than a physical, process?
6. Why cannot all meat be coloured? What changes are involved in meat colour in general?
7. Indicate the qualities that are necessary for the artificial flavouring of:
    (a) ice cream;
    (b) carbonated beverages.
8. How is the flavour aspect of confectionery goods monitored in a quality control laboratory?

7 Chemical kinetics

LAW OF MASS ACTION

Chemical reactions are based upon the kinetic theory of matter. When, for example, ethyl alcohol and acetic acid are mixed together, a reaction to form ester and water commences. This will continue until a point is reached where there is no apparent change, and the concentrations of reactants and products remain constant. This point is known as the *equilibrium position*. If ester and water are brought together, hydrolysis occurs until a similar equilibrium position is reached. In these reversible reactions, increasing the concentration of one of the reactants will speed up the rate at which equilibrium is reached, but will not influence the final balance point.

Guldberg and Waage in 1867 formulated their law of mass action to explain the effects of concentration in a reversible chemical system. The law states that the rate at which any chemical action is proceeding at any instant is proportional to the active masses of the reacting substances at that instant.

Consider a system $A + B \rightleftharpoons C + D$. From the law, the rate at which A and B react to form C and D depends upon the product of the concentrations of A and B. The term 'active mass' refers to the numbers of gramme moles of substance in 1 litre and the concentrations of all reactants is placed in square brackets.

$$\therefore \text{ rate of disappearance of } A \text{ and } B = k_1[A][B]$$
$$\text{and rate of disappearance of } C \text{ and } D = k_2[C][D]$$

k_1 and k_2 are the velocity constants for the forward and back reactions. At equilibrium, these rates are equal and

$$k_1[A][B] = k_2[C][D]$$

The overall equilibrium constant K is found by dividing k_1 by k_2.

$$\therefore K = \frac{k_1}{k_2} = \frac{[C][D]}{[A][B]}$$

Ester formation yield can therefore be based quantitatively providing the value of K can be determined. To determine K, known concentrations of alcohol and acid can be placed in a sealed tube for a few hours in hot water, when a balance point will be achieved. When the tube is broken into cold water the unused acid can be estimated by titration using alkali of known normality and phenolphthalein indicator.

Example
When equimolecular portions of ethyl alcohol and acetic acid are mixed until equilibrium is reached, the amount of unused acid is $\frac{1}{3}$ g/mol. Two-thirds of the acid has therefore been esterified and an equal proportion of the alcohol has reacted with the acid. Thus:

$$CH_3COOH + C_2H_5OH \rightleftharpoons CH_3COOC_2H_5 + H_2O$$

g/mol at equilibrium $\frac{1}{3}$ $\frac{1}{3}$ $\frac{2}{3}$ $\frac{2}{3}$

From the law of mass action the equilibrium constant K is given by:

$$K = \frac{[\text{ester}][\text{water}]}{[\text{acid}][\text{alcohol}]}$$

$$\therefore K = \frac{(\frac{2}{3})^2}{(\frac{1}{3})^2} = 4.$$

The value of K should remain constant no matter what proportions of acid and alcohol are used. By using different amounts of alcohol/g molecule of acid the amount of ester produced can be obtained from the equation.

In the production of esters for synthetic flavourings, the economics of alcohol usage must be balanced against a time factor and the use of sulphuric acid as a catalyst. The law of mass action enables the best conditions to be theoretically calculated before pilot plant scale trials, followed by bulk production, are performed.

When, in the forward or backward reaction, two molecules of one substance take part, the rate of reaction depends on the square of the concentration of that substance. The Contact process for producing sulphuric acid depends upon the interaction of sulphur dioxide and oxygen molecules.

$$2SO_2 + O_2 \rightleftharpoons 2SO_3$$

$$\therefore K = \frac{[SO_3]^2}{[SO_2]^2[O_2]}$$

In the reaction, there are not only double the number of sulphur dioxide molecules to collide with an oxygen molecule but double the number of sulphur dioxide molecules to be collided with. The dissociation rate of sulphur trioxide is also proportional to the square of its concentration.

CONDITIONS AFFECTING EQUILIBRIA

In addition to reactant concentration, there are five other factors which can influence a chemical change.

1. The physical states

Normally, gases interact more quickly than do liquids, and reactions between solids are slower still. Precipitation is instantaneous in aqueous conditions as the ions move rapidly in a liquid medium. The reactions of qualitative and quantitative (volumetric) inorganic analysis are rapid between ions of electrovalent compounds. With solids, the finer the particle size the more vigorous the reaction. Even lead burns in oxygen if in minute particles as a large surface area is exposed for a given mass.

2. Light

Silver salts are affected by light, which provides energy for their reactions. Hydrogen will combine easily with chlorine in bright sunlight. The synthesis of carbohydrate in green plants depends upon sunlight energy. In the chlorination of toluene (methyl benzene) light is one of the factors which decides where the chlorine will go. It has two places: (*a*) side chain-benzyl chloride; and (*b*) nucleus—chlorotoluene. Benzyl chloride is the compound produced.

3. Temperature

The rate of a chemical reaction is always increased by using a higher temperature. This is expected, as the collision factor of molecules accelerates. In the case of a balanced reaction the final position of equilibrium can be affected. The esterification of ethyl alcohol by acetic acid is one of the few chemical reactions in which heat is neither absorbed nor evolved. In the estimation of exalate in food samples, the sample is warmed to 60°C before running in the potassium permanganate. At room temperature, the reaction rate is far too slow for a practical result.

4. Pressure

Gases only are affected as pressure has negligible effect upon the volume of liquids and solids. An increase of pressure is equivalent to an increase in the effective concentration of the reactant. Hence, increased pressure means increased reaction rate. Increased pressure will favour that reaction which is accompanied by a decrease of volume. Examples (with a relevant food application in brackets) include:

$$N_2 + 3H_2 \rightleftharpoons 2NH_3 \text{ (detergent molecule)}$$
$$2SO_2 + O_2 \rightleftharpoons 2SO_3 \text{ (sulphuric acid precursor)}$$
$$3O_2 \rightleftharpoons 2O_3 \text{ (ozonised oxygen, bactericidal influence)}$$

Decreased pressure always favours the product formed with volume increase. Examples include:

$$PCl_5 \rightleftharpoons PCl_3 + Cl_2 \text{ (chlorinating agent-pesticides)}$$
$$N_2O_4 \rightleftharpoons 2NO_2 \text{ (flour bleach)}$$

If there is no volume change then pressure will have no effect on the equilibrium point; it will only speed up the rate at which equilibrium is obtained.

$$H_2 + I_2 \rightleftharpoons 2HI \text{ (iodine-goitrogenic element)}$$
$$N_2 + O_2 \rightleftharpoons 2NO \text{ (meat colour fixation)}$$

5. Catalysts

Catalysis is concerned with chemical kinetics. Catalytic actions are *homogeneous* when both catalyst and reactants are in the same physical state, for example, manganese dioxide in the preparation of oxygen from potassium chlorate. In food catalysis, the action is *heterogeneous*, characterised by different physical states, for example, iron as the catalyst in the Haber synthesis process for ammonia.

Catalysts may alter the reaction velocity but never the final position of equilibrium in a balanced reaction (sulphuric acid in ester preparation). Only a small amount is required (nickel in oil hydrogenation to produce fats) and they are usually specific in character, catalysing certain actions (maltase breaks down maltose to glucose). Catalysts are normally positive, that is, they increase a reaction speed. However, a number are negative and decrease the reaction rate. Hydrogen peroxide is unstable but its decomposition rate may be decreased by traces of acids. Substances which improve the efficiency of a catalyst are called *promotors*. Iron is improved by small amounts of aluminium oxide. Catalyst poisons or *inhibitors* reduce a catalyst efficiency. These inhibitors include hydrogen sulphide and hydrogen cyanide which are also poisonous to man.

In some reactions, one of the products acts as a catalyst. The production of manganese ions from a permanganate titration accelerates the reaction. To this type of reaction the term *autocatalysis* has been given.

Water itself acts as a catalyst in many reactions. Ammonia and hydrogen chloride will not react when perfectly dry and it appears necessary for traces of moisture to be present before many gaseous reactions can occur. Traces of cobalt ions appear to be necessary for certain plant processes and chlorophyll cannot form in the absence of iron, although the metal is not a constituent. Cobalt ion has also been held responsible for the decrease in efficiency of bleaching powder; oxygen is released and the powder becomes sluggish in gas release when acidified.

Enzymes are perhaps the most important catalysts affecting food changes and these are dealt with as they occur in the text.

LE CHATELIER'S THEOREM

A large number of balanced reactions can have their equilibrium point altered when the change is one of temperature. Van't Hoff's law of mobile equilibrium summarises the direction of the change. The law states that, if a system is in equilibrium, raising the temperature will favour that reaction which is accompanied by the absorption of heat, while lowering the temperature will favour that reaction which is accompanied by the evolution of heat.

It therefore follows that endothermic actions are favoured by temperature increase and exothermic actions respond well to a fall of temperature. The change in equilibrium position with change of conditions can be forecast qualitatively by application of Le Chatelier's principle or theorem. This states that if the conditions of a system, initially at equilibrium, are changed, the equilibrium will shift in such a direction as to tend to restore the original conditions.

The vast raw-material requirements of the food industry must be produced as cheaply as possible. Quicklime, used for carcass disposal after foot and mouth disease, and nitric oxide needed for nitric acid production of fertilisers, are both obtained by endothermic actions.

$$CaCO_3 \rightleftharpoons CaO + CO_2 - cals$$
$$N_2 + O_2 \rightleftharpoons 2NO - cals$$

Accordingly, an increase in temperature will, in these reactions, move the point of equilibrium to the right, whilst a decrease in temperature will move it to the left.

The preparation of ammonia is exothermic and is favoured by the lowering of temperature. It must be borne in mind that the incorporation of a catalyst also assists, and a practical minimum temperature of about 200°C will produce ammonia in a reasonable period of time.

Le Chatelier's theorem also explains the effects of pressure upon a system in equilibrium. Pressure favours ammonia production because the system moves in such a way as to oppose the effect of the pressure (constraint) by reducing its volume.

PRINCIPLES OF EXTRACTION AND DISTILLATION

In general, organic compounds are more soluble in organic solvents than in water. The basis of the extraction of fat from food complexes using a Soxhlet extraction unit (Fig. 17) is a direct application of the partition coefficient, as discussed in Chapter 5.

The distillation of essential oils using steam produces a water/oil condensate; chloroform or ether shaken with the mixture will extract most of the oil fraction. On the manufacturing scale, penicillin is obtained from an aqueous solution by shaking with chloroform.

In all such cases, successive small amounts of the extracting liquid are

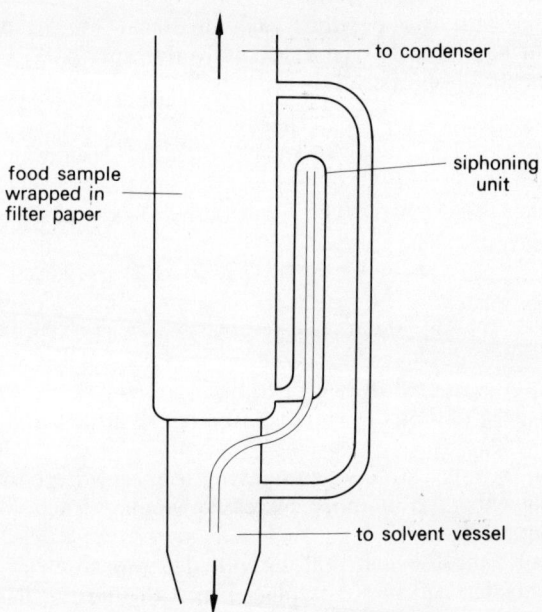

food sample
wrapped in
filter paper

to condenser

siphoning
unit

to solvent vessel

Fig. 17. Soxhlet extraction unit.

used instead of one large amount in one extraction only. The following calculation serves to show that such a procedure gives a better yield of product.

Example

An aqueous solution contains 10 g of essential oil/m^3. When 1 m^3 of the solution is shaken vigorously with 100 cm^3 of ether, 6 g of oil are extracted. How much more of the oil would be extracted from the aqueous solution by: (*a*) a further 100 cm^3 of ether; and (*b*) two extractions each of 50 cm^3 of ether?

Since, initially, 6 g of oil dissolve in the ether, 4 g remain in the aqueous layer.

∴ the partition coefficient of oil between ether and water

$$= \frac{\text{concentration in ether}}{\text{concentration in water}} = \frac{6}{4} \quad \text{or} \quad \frac{3}{2}.$$

(*a*) Extraction with a further 100 cm^3 will remove

$$\frac{4}{1} \times \frac{3}{5} = 2.4 \text{ g}.$$

∴ 1.6 g remains of the original 10 g in the aqueous layer.

(b) Extraction with two portions each of 50 cm³ means the partition coefficient has to be halved as we are only using half as much ether for each extraction operation.

$$\therefore \text{ new partition coefficient} = \frac{3}{4}$$

$$\therefore \text{ first 50 cm}^3 \text{ will remove } \frac{4}{1} \times \frac{3}{7} = 1.71 \text{ g}$$

$$\therefore 4 - 1.71 \text{ g} = 2.29 \text{ g of oil left}$$

$$\therefore \text{ second 50 cm}^3 \text{ will remove } \frac{2.29}{1} \times \frac{3}{7} = 0.98 \text{ g}$$

Total extracted in two operations $= 2.69$ g
$\therefore$ 1.31 g remains of original 10 g in the aqueous layer.

Obviously, in bulk extractions involving tonnes of expensive essential oil the saving with two or more successive small extractions is of considerable magnitude.

Organic substances which boil without decomposing can be purified by distillation. The substance is placed in a distillation flask which is connected to a condenser, the neck of the flask being closed with a cork through which a thermometer passes. The bulb of the thermometer is placed just below the opening of the side tube, and a few pieces of porous pot are put in the distillation flask to prevent 'bumping' or sudden ebullition. In the case of liquids which boil at temperatures above about 130°C, a long glass tube without a water jacket is used (air condenser) instead of an ordinary condenser which is apt to crack. The apparatus is illustrated in Fig. 18.

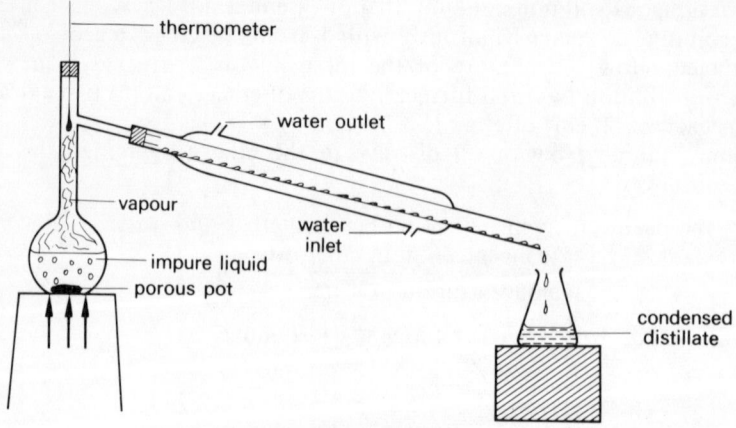

Fig. 18. Distillation of organic liquids.

When the compound to be purified contains only non-volatile impurities, the temperature indicated by the thermometer rises very rapidly as soon as the liquid begins to boil, but then remains practically constant until almost the whole has distilled. If the distillate is transferred to a clean flask, and redistilled, it should boil at a constant temperature providing the external pressure remains the same. It is possible to ensure good separation from any dissolved solute. Water can be recovered free from salt in the distillation of sea-water, though as yet the process is not an economical large-scale proposition.

FRACTIONAL

Providing the boiling points of two liquids differ by at least 20–30°C, it is possible to obtain two distinct distillates from a mixture which contains only a small percentage of the other liquid as an impurity. An example would be the separation of benzene (b.p. 81°C) and xylene (b.p. 140°C). With liquid mixtures whose boiling points differ by between 5 and 10°C the distillation apparatus has to be refined by the introduction of a *fractioning column*. The column is high and consists of hundreds of condensation points inside the main framework, each point being very close to its neighbour. On pilot scale trials involving flavours, heights of 6–8 metres are not uncommon.

The first vapour coming off the liquid condenses at the bottom of the fractioning column. As the liquid so formed flows down the column, it meets more vapour and is redistilled. The process is repeated again and again and thus virtually amounts to a series of separate distillations. By the time vapour reaches the top of the fractioning column to come over into the condenser, it is rich in the most volatile component.

STEAM

Certain compounds will decompose if high temperatures are used in an attempt to isolate them. Fruit flavour components may oxidise. The extraction of natural flavouring components can be achieved without decomposition if steam is passed into a hot aqueous emulsion and the distillate collected in the usual manner. Figure 19 shows the laboratory equivalent of bulk distillation. The method is only applicable to those substances which are volatile in steam and are not decomposed by boiling water.

The ratio of oil in the distillate is obtained from the following formula:

$$\frac{w_1}{w_2} = \frac{\text{molecular weight of water} \times \text{vapour pressure of water}}{\text{molecular weight of oil} \times \text{vapour pressure of oil}}$$

w_1 = weight of water and w_2 = weight of oil

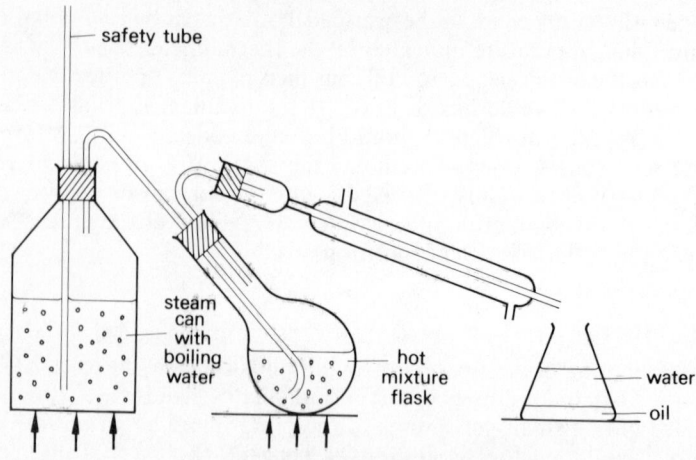

Fig. 19.   Steam distillation technique.

By observing the respective volumes of oil and water in the distillate and using the specific gravities of both liquids, the appropriate weight (or more strictly, mass) can be calculated. In a mixture of two immiscible liquids, each constituent exerts the vapour pressure of the pure liquid at the same temperature. The combined vapour pressures will equal the atmospheric pressure of the day and the individual saturated vapour pressure of water at the distillation temperature can be obtained from tables. It follows that in all cases the distillation temperature of the mixture at normal pressure is below 100°C.

As an illustration, the following proportions of oil and water can be expected in the following mixtures:

Oil	Water
Aniline (1)	Water (2.6)
Bromobenzene (1)	Water (0.62)
Nitrobenzene (1)	Water (3.97)
Ortho nitrophenol (1)	Water (7.16)

Even though the saturated vapour pressure of the oil is low compared to water at the distillation temperature, the molecular weight of the oil is much greater than that of the water. The result is that the distillate contains oil in economic proportions.

Example

An essential oil component has a molecular weight of 128 and distils in steam at 98.3°C under a pressure of 753 mm. The vapour pressure of water at this temperature is 715 mm. Calculate the proportion of oil in the distillate.

$$\frac{w_1}{w_2} = \frac{\text{molecular weight of water} \times \text{vapour pressure of water}}{\text{molecular weight of oil} \times \text{vapour pressure of oil}}$$

$$\frac{w_1}{1} = \frac{18 \times 715}{128 \times 38}$$

$$4864\, w_1 = 12\,870$$

$$w_1 = 2.646$$

Ratio of oil to water is $1:2.646$

$$\therefore \text{ percentage of oil in the distillate} = \frac{1 \times 100}{3.646} = 27.43.$$

Providing only one unknown is present in the equation the molecular weight of the organic constituent can be determined if all other data are given.

REFLUX AND HYDROLYSIS

The preparation of soap from oils or fats demands vigorous treatment with hot alkaline solutions. Under these conditions the triglycerides, of which tristearin is a typical example, are broken down into glycerol and the appropriate salt of stearic acid.

$$(C_{17}H_{35}COO)_3C_3H_5 + 3NaOH = 3C_{17}H_{35}COONa + C_3H_5(OH)_3$$

glyceryl stearate · sodium stearate · glycerol

To provide the continuous intimate contact between the oil and water phases, a normal water condenser is placed vertically above the oil/caustic mixture so that the vapour from the boiling liquid is not lost, but condenses back to liquid and is thus continuously re-cycled. This technique is known as *reflux* and is applied extensively to reactions needing maximum contact conditions.

The reverse process of esterification is hydrolysis. In this reaction, water acts chemically to break down esters into the original acid and alcohol. Usually, weak acids or weak alkali conditions are employed (in association) along with a high temperature. If alkali is used, particularly with the glyceryl esters, a soap and glycerol are produced. The term saponification describes this special type of hydrolysis.

Reflux conditions are used to ensure adequate contact of the reactants. Starch hydrolysis, to produce sugar, usually incorporates acid conditions. Hydration should not be confused with hydrolysis. Hydration will not alter the chemical nature of a compound. Gelatin hydration incorporates increased volume, whereas gelatin hydrolysis produces amino acids.

VACUUM

By lowering the pressure above a liquid, its boiling point falls. This explains the advantage of distillation under reduced pressure, sometimes called vacuum distillation. When a liquid boils at an inconveniently high temperature, or decomposes when distilled at atmospheric pressure, the method of reduced pressure may be applied.

The removal of water from food is desirable on the grounds of preservation and economics of handling. By freezing food, so that the water it contains is turned into ice, placing it in a chamber under a high vacuum and applying a little heat, the ice will sublime. Sublimation is a change of state from solid to gas without the intermediate liquid phase, e.g. iodine and ammonium chloride.

Food which has undergone freezing, followed by sublimation of the ice, will remain stable as long as it is kept dry and protected from the air. By contrast, frozen food is only preserved at low temperatures and this is not suitable for all types of food. The process of 'accelerated freeze-drying' (A.F.D.) has been successfully applied to shrimps, fruit, vegetables and meat. In the case of meat, accelerated freeze-drying is limited, in that it will only work successfully upon thinnish slices. Pork can be treated successfully if it is frozen after being boned out and the loins cut into chops about 1.5 cm thick. When these are placed in the A.F.D. chamber the ice is successfully sublimed, whereas failure results with legs and fillets.

A.F.D. involves elaborate equipment, which is expensive, and needs trained personnel to operate it. However, it is feasible for food items such as shellfish.

Natural fruit essences can be concentrated by boiling water off under vacuum. At the low temperatures involved there is little or no deterioration of original flavour. In the production of condensed milk from sugar and milk, surplus water is removed under a vacuum of about 71 cm of mercury at a temperature of 50–55°C.

Any process demanding water removal from food nutrients is always more successful under vacuum as the denaturation will be at a minimum. However, it is expensive compared to water removal at atmospheric pressure.

EXERCISE 7

1. By applying the Law of Mass Action, deduce what would be the theoretical yield of ethyl acetate from 2 g molecules of ethyl alcohol and 1 g molecule of acetic acid.
2. State Le Chatelier's principle and illustrate its application to physical and chemical changes in food reactions.
3. 7.5 g of ethyl alcohol are mixed with 15.0 g of acetic acid and the mixture is allowed to come to equilibrium at constant temperature. It is found that at equilibrium 7.2 g of acetic acid remains. Calculate the value of the equilibrium constant to one decimal place.
4. State three characteristics of a catalyst. Outline different manufacturing processes of food interest, including one where iron and one where nickel is used as a catalyst.

5. At 98.7°C a mixture of the vapours of an essential oil constituent and water was recovered from a steam distillation. The vapour pressure of water at this temperature is 725.5 mm and the pressure of the mixed vapours was 758.9 mm. Calculate the molecular weight of the oil fraction if the mixture obtained from the distillation contained 23.5 per cent by weight of the oil.

6. A substance is twice as soluble in ether as it is in water, the molecular weight being the same in both solvents. Compare the quantities extracted on a percentage basis from 30 cm³ of aqueous solution:
    (*a*) by 30 cm³ of ether in a single operation; and
    (*b*) by two successive applications of 15 cm³ of ether.

7. What applications are there in a quality control laboratory for food testing using:
    (*a*) reflux;
    (*b*) fractional distillation; and
    (*c*) hydrolysis techniques.

8. Write an account of vacuum distillation in the production of:
    (*a*) a milk product; *or*
    (*b*) a meat product.

8 Physical methods in food analysis

POLARIMETRY AND REFRACTOMETRY

A ray of light can be envisaged as a bundle of waves with its vibratory motions distributed in all directions. If the light is passed through certain minerals, for example, tourmalin or calcite, it emerges vibrating in *one* plane only and is said to be plane polarised. By cementing the two halves of a calcite crystal with Canada balsam, a Nicol prism is produced. When a second Nicol prism is placed in the emergent beam, it will pass only that component of the light vibrating parallel to its axis. It therefore follows that in one position virtually all the radiation will pass through, whereas turning the second prism (the analyser) through a 90° angle will cut the radiation to zero. Two Nicol prisms with their axes perpendicular to one another so that the radiation will be totally absorbed are termed crossed Nicols. Figure 20 illustrates the points discussed.

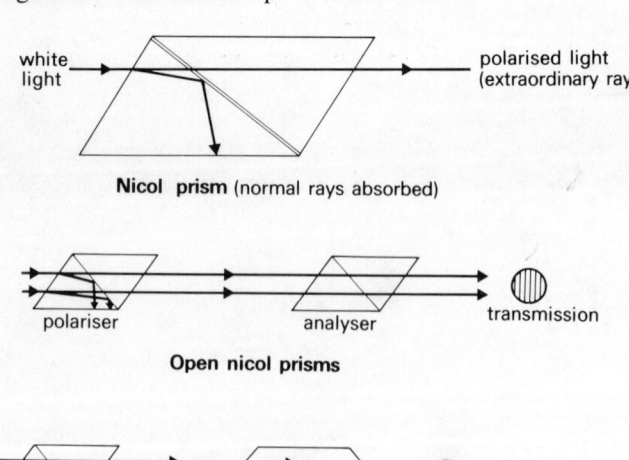

Fig. 20. Polarimetry.

As we have seen, optical isomerism is only possible if the substance contains at least one asymmetric carbon atom. The polarimeter and the saccharimeter are described in detail in Chapter 11.

Nutrients and food additives which have the property of rotating polarised light to the right are dextrorotatory and, conventionally, are said to have a positive rotation. Similarly, laevorotatory or negative rotation is to the left.

The extent to which the plane is rotated by an active substance in solution depends upon the concentration of the substance, the length of column of solution through which the radiation passes, the rotating power of the substance, the temperature and the wavelength of the radiation employed. All these factors are expressed as the specific rotatory power (S.R.P.) of the substance. The identification of proteins and sugars and also their concentration will be related to their specific rotatory powers.

Every transparent or translucent substance will have its own refractive index and this can be used to identify it. For precise work the Abbé refractometer is preferred to the pocket refractometer. The Abbé has a range of refractive indices from 1.3 to 1.7. (The index for pure water, standard reference point for many solutions, is 1.33). The refractometer consists essentially of two glass prisms, adjusting knobs and a telescope provided with cross wires. The sample is placed between the prisms, and the prism box fastened. The telescope is adjusted so that the cross wires are distinct and in focus. By rotating the adjustment knob, a field of vision is obtained in which the lower part is dark and the upper part light. Using the fine adjustment knob, it is possible to position the boundary line at the intersection of the cross wires in the telescope and hence read off the percentage of soluble solids, and also the pure refractive index of the sample under test.

The refractive index of fats and fatty materials will increase with increasing chain length in fatty acids and esters. Fractions obtained in fatty acid ester fractionation can thus be identified. The refractive index also increases with increasing unsaturation. It can act as a control in the hydrogenation of fats and produce samples of various degrees of viscosity and aeration ability. As temperature increases the refractive index decreases, whilst the change in iodine value and refractive index is nearly linear.

Free fat in spray dried whole milk powders can be determined by refractometer measurements. Di-chloronaphthalene acts as a suitable solvent and the filtered extract provides a refractive index which can be converted directly, or indirectly by means of a calibration curve, into the content of free fat. The procedure is simple, only 15 minutes being required to give a result accurate enough for quality control purposes. A deviation in results of only 0.18 can be expected. Full cream milk should not contain less than 26 per cent fat.

COLORIMETRY

Reference has been made previously on the use of colorimetry to determine caramel strength and the potency of a liquid brine cure for meat. Colorimetric determinations have been found to be very valuable for the quantitative determinations of small amounts of a constituent. For larger amounts, gravimetric or volumetric methods are most commonly used, since colorimetric methods are usually less accurate.

Colorimetric methods are more rapid than the gravimetric and volumetric methods available for the same determination. Ethylene, when applied to citrus fruit, will reduce the chlorophyll content of the citrus peel and produce a substantial increased in carotenoid pigments with consequently improved fruit colour. The ethylene-induced carotenoid increase is quite sensitive to temperature and is inhibited at 30°C and above. These changes can be followed in minute detail using matching colorimetric tubes.

Nitrite in brine may result from bacterial action upon nitrates. The permanent pink colour of cured meats is an indication of a preservative action. Nitrite is toxic and the amount added is controlled by food regulations.

A blank solution, and one containing the brine under test, are placed in the separate matching tubes in the Nessleriser. By rotating a colour disc (marked in p.p.m. of nitrite from 0.05 to 1.00) until the blank matches with the brine containing Greiss–Ilosvay reagent, a figure for a known dilution can be obtained. The reagents are: (*a*) sulphanilic acid; and (*b*) Cleves acid. Due to the nitrite content of the brine, the sulphanilic acid is diazotised and coupled with the Cleves acid to produce a pink coloration which is easily compared against the colours on the disc.

Milk freshness can be estimated by the Resazurin test. In this, a blue dye is incubated with a portion of the milk at 37°C for 1 hour. The reducing action of the micro-organisms causes colour changes. These changes can be observed directly, or with the aid of a Lovibond comparator which incorporates a matching colour disc to be assessed against the incubated milk sample. The significance of the matched colour disc is as follows.

Colour after incubation	Disc No.	Quality of milk
Blue	6	Satisfactory
Lilac	5	Fairly satisfactory
Mauve	4	Fairly satisfactory
Pink-mauve	3	Unsatisfactory
Mauve-pink	2	Unsatisfactory
Pink	1	Unsatisfactory
Colourless	0	Very poor

A Lovibond comparator is also used to estimate the whiteness of milk against a cream colour standard when produced as a spray dried powder. In the manufacture of dried milk powder using a hot roller drying technique, a certain amount of caramelisation of sugar occurs. These particles will tend to colour the powder and are removed by a sieving process. The caramelised portion is used as animal feed.

CHROMATOGRAPHY

A protein molecule, when subjected to hydrolysis using dilute mineral acid, will produce a complex mixture of amino acids. These acids have very similar chemical and physical properties and it would be difficult to separate them by a method such as fractional crystallisation. It is, however, possible to achieve a rapid separation using the technique of chromatography, whereby the separation of a mixture of solutes is achieved by utilising their different choice patterns between two solvent molecules. Usually, one solvent is stationary whilst the other carries the mixture along with it.

Tswett may be regarded as the pioneer of chromatography. He prepared a long vertical glass column and packed it with finely divided chalk to act as a stationary porous medium. When a solution of green plant pigments was applied to the top of the column and washed down with light petroleum, a series of coloured bands appeared, while solvent left by a hole at the bottom of the column. These coloured bands occupied different horizontal positions in the column. The term chromatography or 'colour writing' was coined to describe the phenomenon. In fact, the separation is not due to colour but to various pigments having different adsorption affinities for the chalk. The least strongly adsorbed constituents move down the column more rapidly than those strongly adsorbed. This experiment is the first demonstration of *adsorption column chromatography*.

It is difficult to obtain a reliable identification of very small quantities of the separated substances using a column and this has led to the introduction of paper chromatography techniques. The principle depends upon the different rates at which compounds in solution will migrate across a sheet of porous paper. The compounds may be colourless but will appear if the sheet, after being dried, is sprayed with a chemical reagent.

Paper chromatography is also described as partition chromatography since each solute makes a choice between the moving solvent and the stationary liquid held in the pores of the paper as water. In partition chromatography, the moving phase may be either liquid or gas. The various possibilities of both adsorption and partition chromatography are therefore:

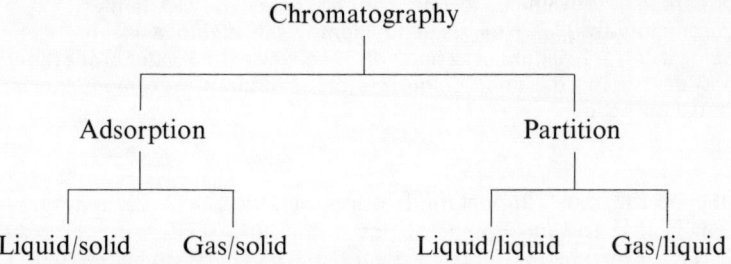

The identification of food colours using paper chromatography may be carried out by extracting the colour from the food and adding a very small amount of water to produce a concentrated solution. A pencilled base line is drawn on to a roll of Whatman filter paper and a small spot of colour is transferred to the centre of the base line using a fine capillary tube. The spot is allowed to dry and the paper is lowered into the solvent mixture (2.5 per cent salt solution and alcohol) so that the base line is about $\frac{1}{4}$ inch above the solvent level. Figure 21 shows the assembled apparatus.

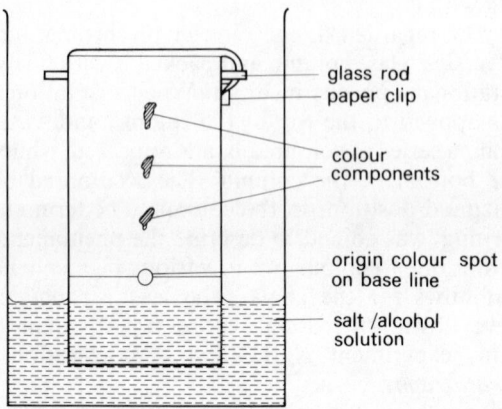

Fig. 21. Ascending paper chromatography

After leaving for 1–2 hours, the individual colour components will appear as separate streaks and the pattern can be compared against permitted food colours. The method described is that of ascending partition chromatography, since the lower edge of the paper strip dips into the solvent.

In descending partition chromatography, the developing solvent is placed in a trough in which the paper is anchored by means of a glass rod. The solvent travels down the paper which has its lower edge serrated so that the solvent will flow uniformly off the paper. For many separations a choice between these two techniques is one of personal preference, as similar results can be obtained with either. Paper chromatography can be used to identify the amino acids in casein using a solvent mixture of 12 butanol : 3 acetic acid : 5 water proportions and developing the dried colourless chromatogram with ninhydrin for visual purposes.

One of the most important features of any given chromatographic system is that the movement of any compound relative to the solvent front is a reproducible and characteristic value. With paper and thin layer chromatography the movement of any compound is known as its R value.

$$R_f \text{ value} = \frac{\text{distance travelled by solute front}}{\text{distance travelled by solvent front}}$$

Paper chromatography is a versatile technique but is confined to cellulosic materials since other media, such as silica gel which is more durable, cannot be made into sheets. This difficulty can be overcome if thin layers of silica gel are supported on a suitable base. The technique is *thin layer chromatography* and excellent separations can be made in 20–40 minutes. Moreover, because of the fine particle size of most thin layer media, improved resolution and compact spots are produced.

Thin layer chromatography is specially suitable for the qualitative and quantitative determination of constituent glycerides in fat analysis. A slurry of silica gel is spread over a glass plate and dried as a thin film. Compounds which have an olefinic linkage are suitable because they are able to form weak co-ordination complexes with silver ions, the complexing strengths of the ions controlling the chromatographic separation of the constituent glycerides. For this purpose, a silver salt is incorporated in the slurry when preparing the plates. By reference to chromatograms of model glycerides, the identity of individual glycerides in a fat sample is established. The dried plates are manipulated as in paper chromatography.

Gas chromatography is the most popular chromatographic method. In gas–liquid chromatography, the solvent is a thin layer of a non-volatile liquid coated on a finely divided, inert solid support. The most popular support is made from celite, which is derived from the silica-containing bodies of microscopic sea creatures.

Separation of a sample mixture is accomplished in the following manner. The sample is introduced from a micro-pipette into a continuous stream of inert carrier gas (usually nitrogen). Each individual component then moves through the column at a rate depending upon its solubility in the liquid phase and its vapour pressure at the particular temperature at which the column is operated. Normally, the individual constituents will emerge from the column in the order of increasing boiling points. The components enter a detector attached to the column where they will generate a minute electrical current related to their concentration and elusion rate. These are amplified and appear on the chart of the recorder as a chromatogram. Figure 22 illustrates diagrammatically the functional parts of a gas chromatograph and also the chromatogram produced from the injection of an ester mixture.

The identification of organo-chloride and organo-phosphorus pesticides such as aldrin, chlordane, endrin and heptachlor is made possible using gas chromatography. Pesticides have a stable structure and may

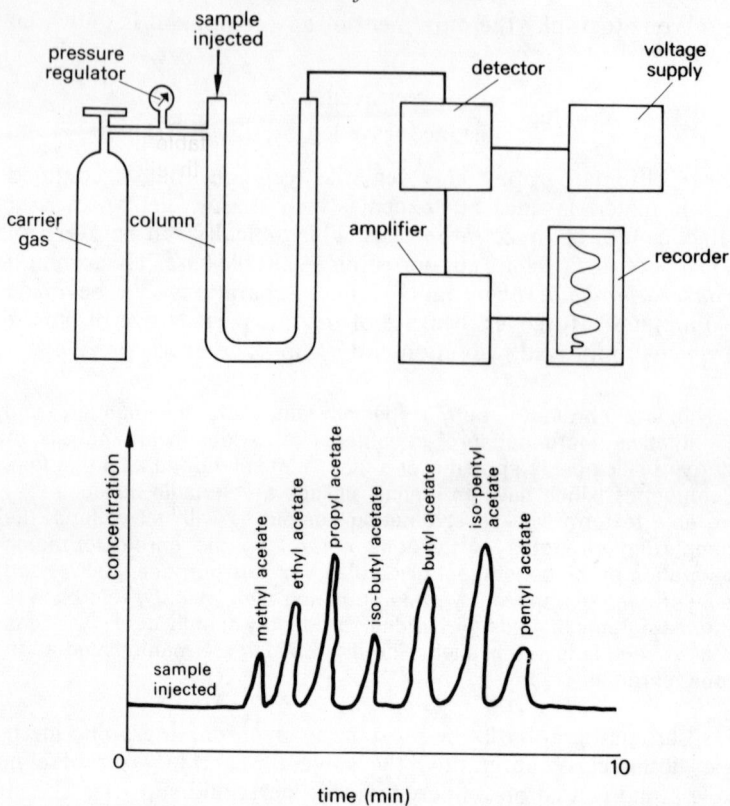

Fig. 22. Gas chromatography.

pass down the food chain from host to host, causing the death of the last. This could be a fox, hawk or other predator, gorged upon the bodies of earlier victims. Farmers in the past have been too liberal with pesticide usage and animal life has suffered accordingly; happily the balance is how being restored as a result of chromatographic investigations. Gas chromatography is also used extensively for the analysis of food oils and flavourings and for following organic reactions.

ELECTROPHORESIS

A feature of colloidal particles is that they are inclined to acquire an electrical charge and, under the influence of an electrical field, they will tend to move towards the oppositely charged electrode. The migration of charged anions or cations under the influence of an electrical field is electrophoresis, and the speed of migration can be used as an identifying feature of individual constituents in a colloidal system.

Electrophoresis consists of applying a differential voltage across a column or strip. The sample constituents will migrate at a speed and direction that is related to the mass and charge of the individual constituents. Proteins can be separated by making use of electrophoresis, different proteins moving with different electrophoretic velocities. It is possible to predict that a given protein, under defined conditions of concentration, pH and ionic strength, will have a net charge, which is defined as the algebraic sum of the charges on the molecule or molecular aggregate. At the isoelectric point no migration is possible, while below it, the protein has a net positive charge and will migrate towards the cathode. Above the isoelectric point it has a net negative charge and will migrate towards the anode.

Milk contains two groups of proteins: (*a*) caseins, and (*b*) serum proteins or whey; examples are lactalbumin and lactoglobulin. The caseins form a closely related series called alpha, beta, gamma, etc. casein which can be separated by electrophoresis. If the protein mixture is applied to a small area of filter paper wetted by a buffer, and current used, after a few hours electrophoresis will have separated the individual caseins. In order to render the colourless components visible, a dye is added which attaches itself to protein but not to cellulose. If the initial protein mixture was applied as a narrow, concentrated streak across the centre of the paper strip, the separated proteins show up as coloured 'bands' on the anode side of the paper (assuming an alkaline buffer was used).

SPECTROSCOPY

The study of wavelengths from a hot body due to radiation is characterised by the term spectra. Elements and compounds all have their individual spectra and millions have been analysed. Because every substance has a unique spectrum, a study of that spectrum enables the composition of the substance to be determined. Spectroscopy is the name given to the exact analysis of mixtures or compounds by a study of their spectra. It is a most exacting science; for example, the presence of less than a nanogram of sodium in a mixture can be detected.

There are three different types of emission spectra: (*a*) line; (*b*) band; and (*c*) continuous. The light emitted by the atoms of a glowing substance, when examined by a spectrometer, produces lines of different wavelengths. (A spectrometer is fully described in chapter 11.) The spectra of hydrogen, helium or sodium produce characteristic features which enable the elements to be easily identified.

Band spectra are obtained from molecules; they consist of a series of bands each sharp at one end but 'fading' at the other. Close scrutiny reveals the bands to be made up of numerous fine lines very close to each other. Nitrogen, oxygen and carbon dioxide will all have their individual band spectra.

Liquids and solids produce a continuous spectra because the atoms and molecules are very close together and each influences the other to produce a complicated pattern. The sun's spectrum is a continuous one and shows a large span of different wavelengths produced by the elements in the sun's atmosphere.

All the spectra just discussed are emission spectra since they have been produced at fairly high temperatures. The identification of nutrients would be impossible using emission spectra techniques; high temperature would decompose the samples and the spectra produced would be that of simple molecules like water, carbon dioxide, ammonia, etc. These molecules must be identified by their *absorption spectra*. For example, if light from a source having a continuous spectrum is examined after it has passed through a sodium flame, the spectrum is found to be crossed by a dark line. This dark line is in the position corresponding to the bright yellow line emission of sodium. The continuous spectrum with the dark line is naturally characteristic of the absorbing substance, in this case sodium, and is known as an absorption spectrum.

Kirchhoff, in the nineteenth century, formulated a law which states that a substance which emits light of a certain wavelength at a given temperature can also absorb light of the same wavelength at that temperature. The spectrophotometer compares the radiations produced, wavelength by wavelength, and since the absorbing nutrient can be at a low temperature its absorption spectrum will serve to identify it.

A spectrophotometric method is available for the detection and determination of casein, particularly in sausages of the Frankfurter and liver sausage types. The procedure is suitable provided that the products have not been sterilised at temperatures of more than 110°C. Lard can be distinguished from hydrogenated vegetable oils by their absorption spectra. The use of infra-red, rather than visible absorption techniques, will prove the presence of trans-fatty acids (geometrical isomers) in freshly rendered beef fat. It can also be used to monitor the trans varieties which may result in the hydrogenation of lard, tallow and other fats. Because of their adverse effect on the assimilation of essential fatty acids, a quantitative determination is necessary.

Chemical compounds can absorb energy in the ultra-violet, visible or infra-red regions for different reasons. For example, proteins absorb energy in the ultra-violet region because of the aromatic amino acids, tryptophan, tyrosine and phenylalanine, whilst those proteins which absorb in the visible region of the spectrum are usually those of the porphyrin type such as haemoglobin and myoglobin.

Food colour may be analysed using a spectrophotometer. In the detection of colour from Annato seeds, an ultra-violet spectrophotometer is used. The method is quicker and more reliable than a colorimetric analysis. An absorption spectrometer can also analyse the metal content of food liquids by reference to standard metal solutions.

There are many commercially available instruments for various types of absorption measurements. Price determines the range of wavelengths covered by an instrument and its degree of resolution. Instruments are available that can scan the infra-red region up to 35 μm and that cover the ultra-violet region down to 220 nm.

MOLECULAR WEIGHT DETERMINATIONS

The molecular weight of gases can be determined from a knowledge of their vapour densities.

Suppose two vessels of equal volume be filled, at the same temperature and pressure, with a gas A and hydrogen, respectively. Then, since the volumes are equal, the number of moles of gas in each volume is, by Avogadro's Law, the same. Let the number be n, and the molecular weights of the two gases be M_x and M_h respectively.

$$\text{Then, the weight of gas A} = M_x \times n,$$
$$\text{Weight of hydrogen} = M_h \times n.$$

By definition:

$$\text{Vapour density of A} = \frac{M_x \times n}{M_h \times n} = \frac{M_x}{M_h}.$$

Since the hydrogen molecule is diatomic:

$$M_h = 2 \text{ (or, more correctly, } 2 \times 1.0080).$$

Hence, since A is any gas, we can write the general relation from which the molecular weight can be obtained as:

$$\text{Vapour density} \times 2 = \text{Molecular weight.}$$

Molecular weight in solution can be determined using four main methods, cryoscopic, ebullioscopic, vapour pressure and osmotic, which will be discussed under their separate headings.

CRYOSCOPIC

It is common practice to produce a freezing mixture by mixing salt and ice together. The freezing point of the mixture is approximately $-20°C$. Blagden found that the lowering of the freezing point of a solvent was proportional to the solute concentration. Beckmann further investigated the subject of freezing point depression or *cryoscopy* and from these results Raoult discovered that equimolecular amounts of different solutes in the same quantity of solvent produced the same depression of freezing point. The laws of Blagden and Raoult only hold true for dilute solutions and if no chemical action between solute and solvent takes place.

For each solvent there is a value, known as the molecular depression or cryoscopic constant. This is the depression of freezing point produced when the molecular weight in grammes of a solute is dissolved in 1000 g solvent. The value can be determined experimentally by measuring the actual depression of a solution of known concentration of a solute of known molecular weight; then, by calculation, the depression which would be caused by dissolving the gramme-molecular weight in 1000 g solvent, is obtained. To measure the depression a Beckmann thermometer must be used. It covers a range of only about 6°C but, using a long stem can measure to .01 of a degree.

The molecular weight of solutes which do not undergo appreciable dissociation, such as organic acids, can be calculated using the following formula.

$$M = w \times \frac{1000}{s} \frac{K}{\Delta t},$$

where M = molecular weight, w = weight of solute, K = cryoscopic constants, s = weight of solvent, and Δt = observed freezing point depression.

Example

A solution of cane sugar containing 12.5 g of sugar in 1000 of water begins to freeze at 0.068°C below zero. For water $K = 1.86/1000$ g. Calculate the molecular weight of cane sugar.

$$M = 12.5 \times \frac{1000}{1000} \times \frac{1.86}{.068} = 341.9$$

The cryoscopic constant value differs for different solvents. A few examples are: water 1.86, benzene 5.12, acetic acid 3.9, and camphor 40.

Camphor is a useful solvent for laboratory work in view of its very large depression.

EBULLIOSCOPIC

In the production of fondant, considerable amounts of sugar are dissolved in water with the result that the boiling point of the syrup is well above 100°C. As with the depression of the freezing point, the extent of elevation may be calculated from the previous laws. For each solvent there is a molecular elevation or ebullioscopic constant, this being the elevation of boiling point caused by dissolving the gramme-molecular weight in 1000 g of solvent. As with the cryoscopic constant, this is obtained for a particular solvent by determining the elevation of the boiling point caused by a known weight of solute, the molecular weight of which is known in the particular solvent.

Some ebullioscopic constants per 1000 g are: water 0.52, acetone 1.72, benzene 2.57, chloroform 3.88, ether 2.12, and ethyl alcohol 1.15.

Example

A solution of 1 g of phenol in 50 cm^3 ether is boiled at a temperature 0.632°C higher than that of pure ether. The molecular elevation constant

of ether is 2.12/1000 g. The density of ether is 0·714 g/cm³.
Calculate the molecular weight of phenol.

$$\text{Weight of ether} = \text{density} \times \text{volume}$$
$$= .714 \times 50 = 35.7 \text{ g}$$

$$M = 1 \times \frac{1000}{35.7} \times \frac{2.12}{.632} = 94.1$$

The same limitations apply to ebullioscopic molecular weight determinations as to their cryoscopic counterparts.

VAPOUR PRESSURE

Every pure solvent has its own vapour pressure. Raoult showed that when a small amount of the liquid was introduced into a barometer tube containing mercury, with a Torricellian vacuum at the top, the solvent vaporised. When equilibrium was reached, the fall in mercury level gave the vapour pressure. The method is not very accurate but it illustrates the principle behind vapour pressure measurements.

The introduction of a non-volatile solute lowers the vapour pressure of a solvent and hence raises its boiling point. Wüllner, in the nineteenth century, showed that the lowering of the vapour pressure of a solvent by the presence of a non-volatile solute varied directly with the concentration. Raoult investigated vapour pressure phenomena using several organic solutes and solvents other than water, for example, ether. He discovered the law which may be stated as: the relative lowering of vapour pressure (that is, lowering divided by vapour pressure of pure solvent) is equal to the ratio of the number of gramme-molecules of solute to the total number of gramme-molecules of solvent and solute.

If P be the pressure of the pure solvent, P' that of the solution, n the number of gramme-molecules of solute and N that of solvent, then:

$$\frac{P - P'}{P} = \frac{n}{N + n}$$

In such a solution, n will be very small compared with N, and an approximation may be used:

$$\frac{P - P'}{P} = \frac{n}{N}$$

If a given weight w of solute of molecular weight m be dissolved in a weight W of solvent of molecular weight M, then substituting:

$$\frac{P - P'}{P} = \frac{w/m}{W/M}$$

If the molecular weight of the solvent M be known, that of the solute in the particular solvent can be obtained.

Example

At 20°C the vapour pressure of ether is 442 mm of mercury. When 6.1 g of a substance was dissolved in 50 g of ether, the vapour pressure fell to 410 mm. What is the molecular weight of the substance?

$$\text{Molecular weight of ether } (C_2H_5)_2O \text{ is } 74 = M$$
$$\therefore N = 50/74$$

Let m = molecular weight of substance

$$\therefore n = 6.1/m$$

$$\frac{P-P'}{P} = \frac{442-410}{442} = \frac{32}{442}$$

$$\frac{n}{N} = \frac{6.1}{m} \Big/ \frac{50}{74} = \frac{6.1 \times 74}{50\,m}$$

$$\therefore m = \frac{6.1 \times 74 \times 442}{50 \times 32} = 125$$

OSMOTIC

Figure 23 shows a small amount of syrup solution, which is of a reasonable concentration, placed in a thistle funnel which has the mouth secured with a piece of parchment to make a 'water-tight' joint. The funnel is then placed in a beaker of distilled water. A rise of liquid occurs in the funnel as water passes into it. Eventually equilibrium is attained when the hydrostatic pressure of the column just balances the tendency of the water to pass through the parchment. Tests show that little or no sugar is found in the water of the beaker.

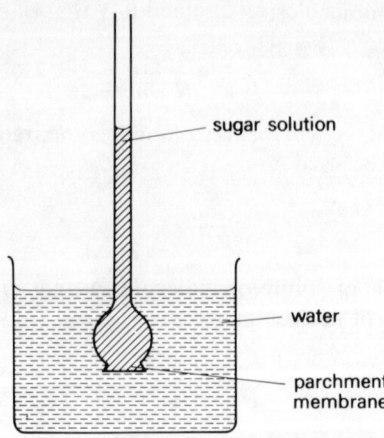

Fig. 23.   Osmotic pressure phenomenon.

The experiment illustrates the phenomenon of osmosis which may be defined as the passage of solvent, but not solute, through a semipermeable membrane. The movement may be from solvent to solution or from a solution of low, to one of higher concentration.

The excess pressure which must be applied to a solution to prevent the passage of solvent into it through a semi-permeable membrane is said to be the osmotic pressure of the solution.

If a salt solution had been placed in the thistle funnel there would have been no rise of liquid, although the salt solution can exert a definite osmotic pressure. The parchment is permeable to sodium and chloride ions, which therefore pass through the membrane until their concentration outside is the same as that inside; that is, diffusion takes place.

Plant and animal membranes are semi-permeable but, like parchment, are not perfect in their selection of solute from solvent molecules. A semi-permeable membrane allows the passage of some substances and not others, and may be regarded as a partition which permits the passage of pure solvent molecules more readily than those of the dissolved substance. Colloidal particles cannot pass through the intestinal wall, which acts as a semi-permeable membrane. Living organisms obtain nourishment by the process of digestion only if the food has been reduced to the size of molecules.

Most living cells, animal and vegetable, are surrounded by membranes. Two hen's eggs of approximately equal size are placed in dilute hydrochloric acid to dissolve off the hard calcium carbonate shell and leave the animal membrane. After washing, one is placed in distilled water and the other in a saturated solution of common salt. In the former, water passes through the membrane into the egg with a consequent increase in size. In the latter, water comes out of the weaker solution of the egg and the egg shrinks. The withdrawal of water from a cell by osmosis is called *plasmolysis*. It can only occur when the cell is surrounded by a solution of greater concentration than that inside it.

The plant spirogyra, which may be found as green slimy threads in stagnant water, consists of chains of cells end to end. If a filament is mounted under the microscope and saturated salt solution drawn over, then the protoplasm which originally filled the cell sets in strands as plasmolysis occurs.

If, on the other hand, a cell is surrounded by a solution less concentrated than that of the cell sap, water flows into the cell by osmosis. A plant obtains water by this means. The delicate cells of the root tips contain outgrowths (root hairs) in intimate contact with the soil. Through these, water passes by osmosis from the soil water of lesser concentration than that of the cell. A plant obtains nitrogen for protein synthesis from the nitrate ions which pass into the root hairs from the soil. Cell membranes appear to possess some selective power which allows the diffusion of certain beneficial substances but prevents the entry of others of no value. The osmotic pressure of a solvent is always increased when a solute dissolves in it.

Solutions of equal osmotic pressure are said to be *isotonic* to each other, since they must have equal particle concentrations. If two solutions have different osmotic pressures, then the one with the higher value is said to be *hypertonic* to the other. The lower value of two given osmotic pressures is always *hypotonic* to the other.

Cake and unwrapped bread should never be placed in the same container. Water will be lost to the cake as the latter has a higher osmotic pressure and will attract water from the bread.

Certain kinds of yeasts are able to withstand a sugar concentration of 80 per cent, such yeasts being said to be *osmophilic*. Occasionally, cake tops will blow due to the yeasts surviving the high osmotic sugar pressure and generating carbon dioxide gas. Osmotic pressure will also produce burst chocolate centres for the same reason. Yeast metabolism is essentially due to sugar molecules passing into the yeast cell through its semi-permeable membrane as water is drawn into the cell by the yeast being hypertonic to its nutrient medium.

The laws of osmotic pressure are analogous to the gas laws. Osmotic pressure is proportional to concentration and absolute temperature. One gmol of a gas at $0°C$ and 760 mm pressure occupies $0.0224 \, m^3$, whilst one gmol of a solute dissolved in $0,0224 \, m^3$ of solution at $0°C$ has an osmotic pressure of 760 mm.

An accurate determination of the osmotic pressure of an unknown protein molecule can be used to determine its molecular weight. Haemoglobin was found by this technique. The method can be equally applied to carbohydrates, as the following calculation illustrates.

Example

A solution of a carbohydrate in water which contains 4 g of solute per $100 \, cm^3$ of solution has an osmotic pressure of 5.3 atm at $15°C$. What is the molecular weight of the carbohydrate?

Imagine the sugar to be a gas in the volume occupied by the solution.

$100 \, cm^3$ of 'gas' at $15°C$ and 4028 mm pressure (5.3 atm) has a volume at N.T.P. of

$$100 \times \frac{273}{288} \times \frac{4028}{760} = 502.4 \, cm^3.$$

At N.T.P. $502.4 \, cm^3$ of 'gas' contains 4 g of carbohydrate.

$$\therefore 0.0224 \, m^3 \text{ at N.T.P. contain } \frac{4 \times 22\,400}{502.4} = 178.3 \, g$$

$$\therefore \text{ the molecular weight is 178.3}$$

The result would indicate a monosaccharide sugar (molecular weight 180).

EXERCISE 8

1. Give two examples of the quality control aspect of refractive index in food problems. How would you assess the quality of cured ham in the laboratory?
2. What advantages are to be found for thin layer chromatography over paper chromotography? Illustrate your answer by reference to the production of any marketable food product.
3. Write brief notes on:
   (a) electrophoresis; and
   (b) gas chromatography.
4. What advantages are offered by a spectrophotometer over a colorimeter in monitoring the quality of meat products?
5. An aqueous solution of a weak organic acid obtained by fat hydrolysis containing $20 \ kg/m^3$ of the acid freezes at $-0.37°C$. $10 \ cm^3$ of the same solution requires $16.6 \ cm^3$ of $1.2 \ \dfrac{N}{10}$ sodium hydroxide for neutralisation. What is the acid's basicity and its molecular weight?

(K for water per 100 g = 18.6)

6. How can the molecular weight of a pure solvent be determined by an ebullioscopic method? What limitations are involved?
7. 0.511 g of a solid food acid dissolved in 40 g of water caused an elevation of the boiling point equal to $0.073°$. Calculate the acid's molecular weight and suggest what the acid might be, allowing for a very small experimental error. The molecular elevation constant for water (100 g) is 5.2.
8. A solution of a liquid raw material has a concentration of $50 \ kg/m^3$. It has an osmotic pressure of 982.6 cm of mercury at $17°C$. Calculate its molecular weight. If the liquid is used in confectionery work as a humectant suggest what it might be.

9 Matter from a physical viewpoint

SOLID PROPERTIES

The characteristic feature of a solid is the close packing of the atoms that compose it. Crystalline structures can be examined by analysing the information obtained from X-ray diffraction patterns formed by the crystal. There are a number of substances which appear to be solids but are actually very viscous liquids. Glass and pitch are in this category but their rate of flow under gravity is so slow that for practical purposes they appear to have fixed dimensions at room temperature.

Firmness of texture is one important property of freshness in many food sources. A simplified instrument, a texturometer, which can exert variable and programmable rates of strain on different food products has been developed for this purpose. These texture measurements, when compared with the results obtained by viscometry, compression and puncturing are invaluable as regards an accurate assessment of the quality of items like processed and molten cheese, yoghourt, raspberries and strawberries.

The mechanical development of dough by the Chorleywood bread process in 1961 has revolutionised bread production in the United Kingdom. Both the starch and gluten of wheat flour have different physical properties which need recognising. Starch absorbs water and eases the problem of even mixing of constituents, whilst the gluten needs energy to enable water to be distributed over it and cause the tightly packed molecules to open out ready for the development process. The gluten molecules are thus mechanically broken in order to build up more complex molecules.

Oxidisers are usually added to neutralise the hydrogen atoms which have been detached. The theoretical energy required to break the molecules is about one-tenth of what in practice is necessary, but this merely means that the chance of hitting in the right place is about one in ten. The energy input has to be controlled and shows as a temperature rise in the dough due to frictional forces, which can be embarrassing in hot climates.

Food rheology is placing at the disposal of the food manufacturer valuable data about the overall excellence or otherwise of his products.

Rheology may be defined as the study of the deformation and flow of matter. Figure 24 shows graphically the information yielded about two types of cherry, sour and sweet, when subjected to an Instron testing machine. The samples are held rigid whilst a pitter is applied to the skin surface, and the force needed to penetrate the skin, and then contact the pit in the cherry, is measured. The height of the second peak represents the force needed to push the pit through the flesh of the cherry and is a measure of flesh firmness. Machines of this nature can be adapted for texture measurements of fruit, vegetables, meat, biscuits, cake and doughs.

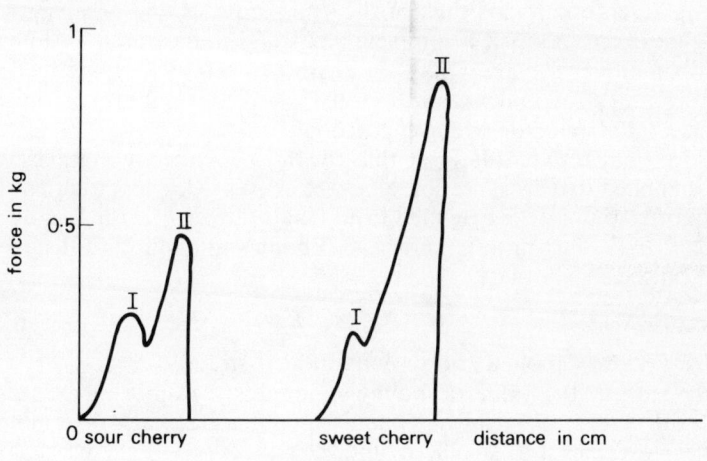

I=pitter penetrating skin

II=increase in force which occurs when the pitter contacts the pit in the cherry

Fig. 24.   Instron testing machine.

The packing of solids also calls for a knowledge of their bulk density. This is not the same as their relative density. The bulk density is always much lower than the unit density because of the large number of spaces between the particles, as the table below indicates:

Item	Unit density	Bulk density
Apples	700–900	530–600
Potato	1080–1170	760–775
Wheat	1410	805

The unit for both density evaluations is kg/m^3.

Bulk density is very important in the preparation of milk powder for babies. There is a danger of underfeeding or overfeeding by the mother who tends to measure the powder needed by spoon rather than weight. A correct bulk density by the manufacturer will ensure that an adequate amount of powder is available for purposes of reconstitution.

TENSILE STRENGTH

The amount of force needed to break a material is a measure of its tensile strength. It is usually measured in force per unit area using newtons per square metre. A qualitative investigation would necessitate having several materials each of the same cross section and length; the force needed to produce a complete break is then measured. Lead has an approximate tensile strength of only $\frac{1}{25}$th of steel. In the construction of holding tanks for milk stainless steel is used, not only for its hygienic surface but also for its strength potential.

Closely allied to tensile strength is the ratio obtained by comparing the stress applied to a cross section of a wire or rod to the longitudinal strain produced. If a body is extended from length L to L+1, the longitudinal strain is 1/L. This ratio is known as Young's modulus and denoted by *E*. It can be shown that:

$$s = e \times E \quad \text{or} \quad E = s/e,$$

where s = stress applied and e = longitudinal strain.

The smaller the value of Young's modulus the more easily does the material deform elastically. Young's modulus has been determined at optimum values for spaghetti and potato crisps. The work can be extended to cover the whole cereal range. Snap or crispness of such products is a criterion of quality.

Low-density polyethylene films used as packaging materials for foodstuffs have been tested for their physical strength after being treated with ionising radiations for pasteurisation purposes. Of the mechanical properties, tensile strength and elongation after break were tested. Both showed a decrease with increasing radiation doses. However, it has been shown that pasteurising doses cause no change in a tested 0.004 cm thick low density polyethylene film which would render it unsuitable for the packing of foodstuffs.

ELASTICITY

Rubber is a typical example of an elastic material. It has the ability to recover its original dimensions when a distorting force is removed. If the forces on a substance are sufficiently large for the deformation to cause a break in the molecular structure of the body or material, it loses its elasticity and the elastic limit is said to have been reached.

Tin may be regarded as an elastic metal and is used as a separator for different kinds of cakes or biscuits on a moving conveyor belt. Eventually, its elastic limit is exceeded and new strips have to be introduced into the assembly line.

A fruit jelly, prior to syneresis (jelly weeping), has elastic properties which characterise it on a quality basis. The elastic nature of gluten is responsible for the final texture of the finished loaf, fat and ascorbic acid will mellow it but potassium bromate will toughen and produce a harder crumb.

A simple but ingenious device for separating good quality cranberries from over-ripe or poor quality fruit is based upon the elastic ability of a good berry. The fruit is allowed to fall in front of a 10 cm barrier, only those berries which clear the obstacle are selected.

Muscle elasticity depends upon the presence of actomyosin, which is responsible for relaxation and contraction in the live animal. Adenosine triphosphate is necessary for muscle relaxation. Contraction results when adenosine triphosphate (ATP) splits into phosphorylated actin (P) and adenosine diphosphate (ADP). Enzymes are responsible for these changes, which are in equilibrium with each other under living conditions. After slaughter the muscles contract and stiffen (rigor mortis), the energy for this contraction coming from the conversion of stored glycogen into lactic acid. At a given pH the enzymes are poisoned and therefore the ADP cannot be converted back to ATP to produce relaxation.

$$\text{ATP} \underset{\text{relaxation}}{\overset{\text{contraction}}{\rightleftharpoons}} \text{ADP} + \text{P}$$

PLASTICITY

A plastic material has the ability to be moulded into shape and retain its new dimensions. At room temperature, only sodium and potassium are truly plastic metals. Aluminium foil can be used as a wrapping material for sealing the juices of cooked joints. Plastic films include polythene as itself or as a laminate with other wrappings. Polystyrene and polypropylene, when warm, can be moulded into any desired shape and are excellent as containers for food; they impart no odour and their surfaces are easily cleaned.

Meat fats are plastic materials, they consist of masses of small crystals in which a small quantity of oil is entrapped. As the solid content increases, the fats become firmer and less plastic. Lard usually has a solid content of 15–25 per cent within the temperature range 20–30°C.

Normally, long plastic ranges are found in mixtures of glycerides of widely differing melting points. Smooth lard has many small crystals which extend its plastic range and produce the characteristically smooth appearance. Lard and meat fat shortenings are plasticised by rapid cooling using a Votator chilling machine. Rapid cooling causes small crystals to be produced and thus a firm product results.

The plastic nature of marzipan and fondant are put to good use in the decorating of cakes. Fat used in cake production should have a plasticity which will enable it to form a thin sheet or layer in a batter and thus retain air bubbles when 'creamed'. Lard would not cream satisfactorily; it has too firm a texture.

ADHESION AND COHESION

The force which holds a postage stamp to an envelope is adhesive whilst the tendency of water and mercury molecules to remain as separate entities is due to their cohesive forces. Both adhesive and cohesive forces depend upon molecular attraction. On closer examination, adhesion is a force between *unrelated* molecules whilst cohesion is between *related* molecules.

Lining papers for cakes are often coated with silicone compounds to prevent part of the cake sticking to the tin, and the use of fluorinated ethylene plastics for non-stick frying pans is well established. In both cases there is considerable lowering of the adhesive force between dissimilar materials.

The adhesion of smooth packages and glutinous substances to conveyor belts has always been a problem. An open weave which allows small pockets of air to be trapped between the belt and the goods facilitates release. However, this is a rough and ready solution which often creates as many problems as it solves. Cleaning is difficult and a high degree of vigilance is needed to ensure that hygienic conditions are maintained. Recently, a range of conveyor belts based on polyester fibre (Terylene) have been produced which, according to the manufacturers, reduce, and in some cases eliminate, the problem of adhesion. The belt is woven in such a way that it is flexible along its length whilst being very stiff across its width. Since polyester is immensely strong, a very thin light belt results which does not stretch, can negotiate small drums and does not wrinkle. The underside of the belt has a polyurethane surface which slides easily over skid plates, resulting in smooth operation. The reduction in drag also bestows the benefit of lower power consumption.

Wheat quality, as determined by the Farinograph, shows that the farinogram curve, which measures the resistance offered by the dough to the mixing blades, is affected considerably by the adhesive forces between the dough and the blades.

The degree of stickiness of boiled sweets and cake batters can be ascertained by measuring the force needed to detach a plunger from the surface of the materials.

Mercury is ideal as a thermometric liquid for several reasons, amongst which is its ability not to wet glass. The cohesive forces between mercury molecules are much higher than the adhesive forces between mercury and glass. A convex meniscus results which is the exact opposite of the concave water meniscus (water wets glass).

LIQUID PROPERTIES

The molecules of a liquid have considerable latitude of movement compared to solids. Liquids also assume the shape of containers into which they are placed. Hydraulic braking systems are efficient because of the incompressibility of a liquid and are to be preferred over solid systems; the latter generate a large amount of heat and would constitute a fire hazard.

One important property of certain liquids is that of thixotrophy. A thixotropic liquid has different rates of flow depending upon a time factor. If left undisturbed, they will flow more slowly as time increases, but, on shaking, the rate of flow returns to its original value. An important consequence of this time dependency is that the testing routine must be strictly timed. In addition, a previous history of the sample as regards its degree of handling must be known. Chocolate can exhibit thixotropic properties. Time and temperature are therefore important in the enrobing process.

DIFFUSION

Because of the lower kinetic energy of liquid molecules, the diffusion of different liquids to form an homogeneous solution is much slower than that of gas mixture. When distilled water is added carefully, using a pipette, to a layer of concentrated potassium permanganate in a beaker, two distinct layers can be produced. Over a period of hours the boundary becomes less clear and eventually a pink solution of even concentration is obtained.

In the meat industry, the rate of curing of primal cuts depends upon the rate of diffusion of the curing ingredients into the tissues. The rate of diffusion depends upon the size of the cut, the amount of fat covering and the temperature. An increased rate of diffusion of curing agents can be achieved with temperatures in the order of 45–50°C. Although the danger of bacterial spoilage may be increased, this principle has received some limited success with bacon and pigsfeet. The bellies, or feet, are held for several hours at this high temperature, until the curing agents have achieved good penetration. These temperatures exceed the maximum temperature limit for most of the spoilage bacteria.

VISCOSITY

When a liquid is vigorously stirred the liquid rotation will be easily visible. However, as soon as the stirring device is removed the friction between liquid molecules rapidly slows any visible movement. This liquid friction, or internal resistance of a liquid to flow, is called viscosity. An increase of temperature will usually produce a faster flow or lower viscosity, while lowering the temperature causes reduced flow or increased viscosity.

The SI unit of viscosity is the poiseuille and is a measure of the resistance offered to flow by applying pressure to the liquid. If a force of 1 N produces a velocity of 1 m/s at the surface of a liquid then the coefficient of viscosity of the liquid equals one poiseuille (Pl). One Pl therefore equals 1 Ns/m^2. Originally, viscosity was measured in centipoises or poises. One poiseuille equals 10 poise or 1 centipoise $= 10^{-3}$ Ns/m^2.

A simple type of viscometer is illustrated by the flow cup. It is useful for measuring the viscosity of liquids, such as chocolate or tomato purees, where the rate of flow is fairly slow. In essence, it comprises a metal cup drawn out at one end to form a fine orifice, or jet, which has a protective sleeve around it. The sample is poured into the cup and the time of efflux of a definite volume of liquid is measured. A range of different orifice diameters are available to cover the needs of a wide variety of liquids.

In practice, in many facets of the food industry, the viscous quality of a raw material is expressed in terms of the time taken for a given volume to flow, or the time taken for some object to fall through it in a closed container. Figure 25 shows some of the various types of viscometer available.

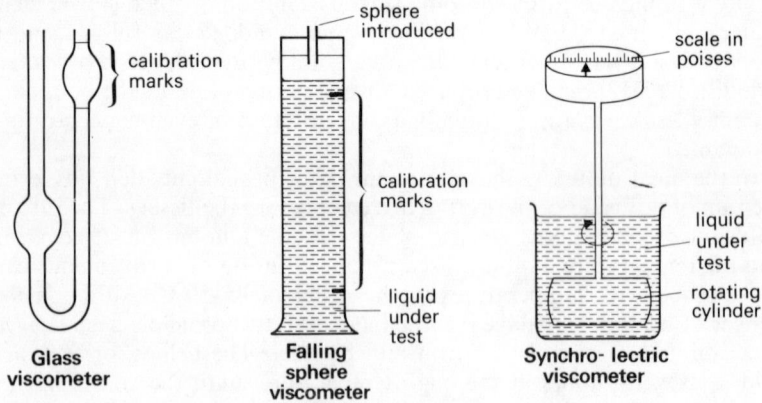

Fig. 25. Viscometers.

The glass viscometer is very economical and requires only the use of a stopwatch in its operation. Initially, the larger lower bulb is filled with the liquid under test. By applying suction to the opening above the calibration marks, liquid is drawn just above the upper graduation mark. The suction is now removed and the time taken for the liquid to pass between the two calibration marks is noted in seconds. An average of three readings will be adequate for the determination of syrup strength. Water is usually taken as a standard for comparison of times.

For liquids with viscosities much higher than water, the falling sphere viscometer is used. Edible oils and glycerol-based liquids are good examples. A sphere commonly used is a steel ball of 1.5 mm diameter. The diameter and distance of fall between the calibration marks must be very large compared with the sphere diameter. Again, the time taken is a measure of the liquid's concentration or purity.

If the liquid is opaque, a metal detector can be employed. The stirrer viscometer is a modified falling sphere type and is used to determine the Hagberg falling number of cereals, for example, flour. In the determination, flour and distilled water in pre-determined amounts are placed in a viscometer tube and shaken twenty times to produce a suspension. A stirrer is now fitted in position and the tube is immersed in boiling water and clamped in position. The time is noted and after 5 seconds the mass is stirred at the rate of 2 stirs/second for 59 seconds (118 stirs). The stirrer is now raised to its full height and allowed to fall until contact is made electronically with a timing device and the time noted in seconds. This time is known as the Hagberg falling number. The falling number value is equal to the stirring time plus the dropping time.

Amylase activity in the starch begins to produce sugar; therefore, the time factor is a measure of the diastatic activity of the flour. A time of up to 150 seconds indicates a very highly diastatic sample with good saccharification; 150–250 seconds shows moderate diastatic action, whilst a time greater than 300 seconds registers low or poor enzymatic activity. Information of this type would indicate what blends of flour could be used to produce various goods, for example, bloomer leaves and plain tea cakes.

The torsion viscometer may be used to compare liquid resistance to flow on a calibrated scale rather than on a time basis. A wire has a suitable weight attached to it, above which is a pointer which can be moved when the wire is twisted to read zero on the scale. If the tension on the wire is removed, the weight will revolve in the liquid under test and its amplitude will indicate the viscosity of the sample. Various weights are available to give a reading within a reasonable time; each liquid will have its own torque reading and a quality correlation established. Rapid readings can be obtained but the apparatus is not portable, samples having to be brought from the factory floor to the test laboratory. The readings are also empirical and are not expressed in SI units.

Industrially, the synchro-lectric viscometer is finding greater use. It is expensive but has the virtue of requiring no timing; it is also portable and gives very quick results. A rotating cylinder of known dimensions is attached by a screw thread to the underside of a scale which is usually calibrated in poises. The cylinder is immersed in the liquid (for instance, a soup), and the cylinder is allowed to rotate electrically. A measure of the liquid's resistance to motion shows on the scale and by using different cylinders a reading can be obtained. The speed of rotation is inversely proportional to the viscosity of the liquid.

Viscosity is of interest to those engaged in the processing and handling of fats where the design of pumps, piping systems and the proper temperatures for handling are involved. Usually, viscosity will increase with the chain length of the fatty acids, and decrease with increasing unsaturation. Butter, cream, or jam for cake fillings are delivered on a commercial scale through polythene pipes. The amount of ingredient added will be related to viscosity and pump pressure.

Under standard conditions every food liquid will have its own viscosity. This will be an important quality factor of foods. Evaluation of this property enables control with raw materials, food products and food at a stage in production where the degree of flow is at an optimum.

SURFACE TENSION

Because of surface tension all fluids attempt to form a minimum surface under all conditions. The minimum surface for a given volume is a sphere. Therefore, when free to do so, liquids assume a spherical shape. Surface tension was formerly measured in dyn/cm, but the SI unit is 1 N/m.

Water has a high surface tension when compared with a large number of organic substances, being approximately three times that of ethyl alcohol at the same temperature. For mercury the figure is about five times that for water under the same conditions, which is in keeping with its strong cohesive forces already mentioned.

Surface tension decreases with an increase in temperature, becoming zero at the critical temperature, that is the temperature at which liquid and saturated vapour have the same density. When a substance is dissolved in a pure liquid, the surface tension of the solution may not be changed; however, it may be raised or it may be lowered.

Aqueous solutions of most electrolytes and some organic compounds slightly elevate the surface tension. Sugar increases the surface tension of water, whilst fats and proteins lower surface tension. The effect is always increased with concentration. Surface tension can be lowered greatly, but it can be raised only slightly. Substances which produce a depression of surface tension will accumulate at the surface. In a gelatin solution, there is more gelatin in the surface layer than in the centre of the solution, whilst in a milk foam the concentration of protein is greater in the foam than in the solution. The incorporation of conditions and additives to produce stable foams or emulsions necessitates a sound appreciation of surface tension phenomena.

HUMIDITY

A study of the amount of water vapour in the atmosphere or its humidity is fundamental to the preparation of certain products and health of personnel in a food environment. Except in cold weather human beings perspire continuously; the perspiration evaporates and draws its latent heat of evaporation from the skin, thus producing a cooling effect. However, if evaporation is too rapid from the skin surface, particularly around the mucous membranes of the mouth and nose, the skin will crack.

The rate at which water evaporates from the skin or a food surface depends upon the pressure of water vapour surrounding it. When the air is fully saturated with water vapour no evaporation is possible. Hygrometry is the study of atmospheric humidity. Hygrometers (not to be confused with hydrometers) are instruments which measure the hygrometric state of the air. Humidity may be expressed as (*a*) absolute or (*b*) relative. The former is only of academic interest in that it expresses on a weight/volume basis the amount of water vapour in a given volume of air. Relative humidity is the ratio between the amount of water in a given volume of air and the amount of water vapour needed to saturate the air at the same temperature.

$$\text{R.H.} = \frac{\text{mass of water vapour in a given volume of air}}{\text{mass of an equal volume of saturated water vapour at the same temperature}}$$

or alternatively

$$\text{R.H.} = \frac{\text{density of water vapour in the atmosphere}}{\text{density of saturated water vapour at the same temperature}}.$$

An unsaturated vapour behaves approximately according to Boyle's Law, and therefore its density is roughly proportional to its pressure. The relative humidity expressed on a percentage basis is:

$$\text{R.H.} = \frac{\text{partial pressure of water vapour present} \times 100}{\text{S.V.P. at temperature of the air}},$$

where S.V.P. stands for saturated vapour pressure.

By allowing water at different temperatures to evaporate in a mercury barometer and measuring the fall of pressure produced at saturation point a table for the S.V.P. of water covering the temperature range 0–100°C can be constructed. Clearly at 100°C the S.V.P. of water equals 760 mm mercury pressure and the liquid will boil. In order to evaluate the numerator in the last formula a dew point hygrometer is used.

The dew point hygrometer consists of a glass tube partially filled with a volatile liquid such as ether. Around the base of the tube is placed a thimble of some shiny metal. Air is blown through and the ether evaporates causing a chilling effect on the thimble. When the temperature is low enough the air in contact with the thimble becomes fully saturated and condensation occurs producing a dull surface. The temperature of S.V.P. is now observed at the dew point.

Example

On a day when the air temperature was 17°C the dew point was 10°C. What was the relative humidity? (Given S.V.P. at 17°C = 14.5 mm of mercury and S.V.P. at 10°C = 9.2 mm of mercury.)

$$R.H. = \frac{S.V.P. \text{ at dew point}}{S.V.P. \text{ at air temperature}} = \frac{9.2}{14.5} = 0.63 \text{ or } 63 \text{ per cent.}$$

In the bakery relative humidity is easily determined using a wet and dry bulb hygrometer. Two ordinary thermometers are mounted in parallel, one having its bulb covered by a piece of muslin which dips into a trough of water. Water is drawn upwards by capillary attraction and the consequent evaporation cools the bulb to an extent dependent upon the dryness of the surrounding air. The rate at which water evaporates from the muslin increases as the relative humidity of the atmosphere falls. The greater the difference in reading of the two thermometers, the less is the relative humidity. By observing the difference, and using the tables provided, the relative humidity is easily obtained. Figure 26 shows the hygrometers discussed and an abbreviated table.

Flour has a moisture content of 11–14 per cent depending upon the relative humidity. The moisture value is of importance in producing dough of satisfactory quality. Salt, sorbitol and sodium lactate absorb moisture readily along with sugar. Dried eggs and milk powder with a 2–3 per cent moisture content are particularly prone to absorbing moisture from the air.

The use of special sealed containers, often containing absorbent salts, to produce a dry atmosphere of very low relative humidity are called for. Fondant is very deliquescent, cakes and bread tend to lose moisture to the air and dough will skin over unless covered with a damp cloth. During normal bulk fermentation the loss of alcohol and water by the dough should be of the order of 3 per cent.

Meat storage should be under optimum relative humidity conditions if weight loss is to be avoided and mould growth inhibited. There is a preferred relative humidity for all basic raw materials. Working personnel usually require a humidity of 55–60 per cent whilst a dough proving chamber needs approximately 90 per cent. In order to meet these requirements humidifiers may be installed to give any desired humidity.

A humidifier blows saturated water vapour into a room of higher temperature. Evaporation quickly takes place and a stable humidity is obtained.

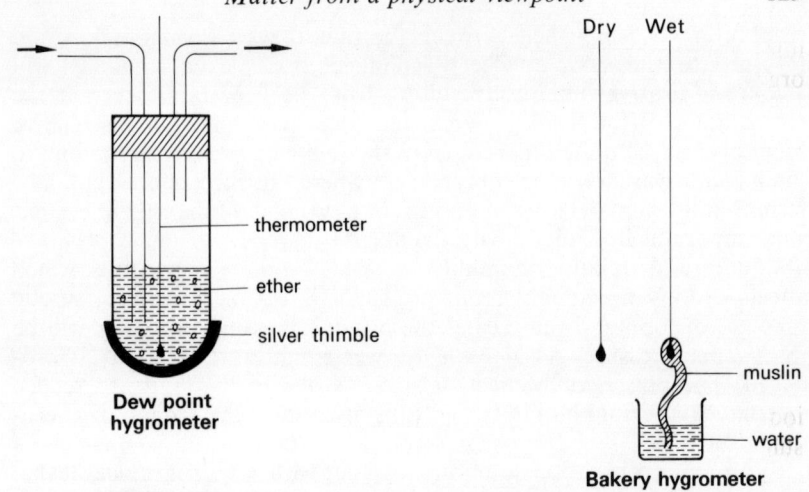

Fig. 26. Hygrometry.

Dry bulb reading °C	Difference (depression of wet-bulb) °C							
	1	2	3	4	5	6	8	10
0	82	65	48	31%				
5	85	72	58	45	32%			
10	88	76	65	54	44	34%		
15	90	80	71	61	52	44	27	12%
20	91	83	74	66	59	51	37	24%
25	92	84	77	70	63	57	44	33%
30	95	86	79	73	67	61	50	39%

Hygrometer table

Example

A raw material testing laboratory is designed to operate at 20°C with 60 per cent relative humidity. From tables the S.V.P. of air at 20°C is 17.5 mm, therefore the vapour pressure required is 60 per cent of this, or

$$\frac{17.5}{1} \times \frac{60}{100} = 10.5 \text{ mm.}$$

This is also the S.V.P. of air at 12°C. Air fully saturated at 12°C is blown from the humidifier into the air at 20°C and produces a humidity of 60 per cent. By using reference tables the humidifier can be adjusted to meet any specific humidity requirement.

A large variety of foodstuffs are subjected to display in disposable 'see-through' packages, usually these are of a plastic origin. When the food in question contains water it will exert its own vapour pressure until equilibrium is established between the rate at which water is lost and returned to the sample. Cheese confined in a plastic container will acquire a relative humidity resulting from its vapour pressure at the temperature of the display. When the package is air-tight this relative humidity will be quite different from the relative humidity pertaining to the air in a warehouse or supermarket where the cheeses are kept. The term 'equilibrium relative humidity' is used to describe this select but very important humidity environment.

Equilibrium relative humidity depends upon temperature which affects various individual vapour pressures. Sugar increases the osmotic pressure of food and the greater the osmotic pressure the lower will be the vapour pressure. As a result a lower equilibrium humidity is produced. The preservative effect of sugar is due to its ability to reduce internal equilibrium humidity and thus discourage the growth of micro-organisms.

Syrups and other food solutions together with solid substances (cake, bread, biscuits) thus produce an equilibrium relative humidity if present in an enclosed space. The proportion of dissolved solids present in the sample has a distinct bearing upon the equilibrium relative humidity.

Mould growth in foodstuffs wrapped or sealed from the atmosphere can be discouraged by observing the equilibrium humidity existing in the container. Mould growth is not extensive below an approximate humidity of 75 per cent. In confectionery goods the normal range of relative humidities may be expected to lie between 65 and 85 per cent. Certain crystals are known to deliquesce (absorb moisture from the air) at different relative humidities and a range can be obtained to cover the percentages met with in finished goods.

For example, laboratory tests at 23°C have shown that potassium iodide will deliquesce at approximately 69 per cent and ammonium sulphate at 78 per cent.

If small holes are cut into a piece of bread or cake and a glass slide placed under them small crystals of the two salts can be added. A glass slide placed over the top to cover the holes will then produce an internal environment and the sample can be incubated at 23°C. Should the potassium iodide liquefy but the ammonium sulphate remain dry then the equilibrium relative humidity must be greater than 69 per cent but less than 78 per cent.

The shelf life of goods is therefore more responsive to control of internal equilibrium relative humidity than external environmental relative humidity, a factor which is sometimes overlooked in the display and rotation of bakery raw materials or finished products.

GAS PROPERTIES

Because their molecules are very far apart gases are able to undergo compression and all gases can be liquefied under appropriate conditions of temperature and pressure. Every gas will have a critical temperature above which it cannot be liquefied by pressure alone. Ammonia has a critical temperature of 130°C which contrasts sharply with -240°C for hydrogen. This explains the relative ease of ammonia liquefaction compared with hydrogen.

Gas production in cured meats may produce trouble especially in the manufacture of fermented sausages. Lactobacilli and more rarely yeasts will produce large amounts of carbon dioxide from the fermentation of added sugar. 'Pin holes' are produced in the sausage which swells and burst casings are often encountered. Gas volume is altered dramatically by temperature changes. The consistent quality demanded of many bread and confectionery products has a direct relationship with temperature and hence gas volume control.

DIFFUSION

At a constant temperature, all gases diffuse at rates which are inversely proportional to the square roots of their densities. This is Graham's Law of diffusion. It follows that the greater the molecular weight of a gas the slower is its rate of diffusion.

Ammonia has a higher rate than carbon dioxide. Leaks on refrigerated vessels using ammonia can be detected quickly using Nessler's reagent which turns brown with 1 p.p.m. of ammonia in the air. At this concentration the odour cannot be detected by most people but the ammonia released could contaminate food because of its high solubility.

Air and steam diffusion raise puff pastry whilst the surface of cracker biscuits can become patterned as gas cracks its way through the outer surface becoming impervious. The rapid diffusion of steam can produce the hard, shiny texture characteristic of Vienna bread.

The packaging of hams and frankfurters under vacuum may result in spoilage. Gas diffuses from the product and is retained by the package resulting in an inflated sample. Because of its low rate of diffusion carbon dioxide can be commercially employed as a 'stunner' for pigs.

The rapid diffusion of air and hydrocarbon gases such as propane and butane demands care if liquefied hydrocarbons are used as a source of heat for cooking meat. On no account should any residual air be in the pipes or an explosive mixture will result.

SOLUBILITY

The solubility of a gas in water normally diminishes as the water temperature rises. An exception is nitrous oxide which is more soluble in warm water than cold. Gases of interest to the food technologist can be classified into three main categories: those that are fairly insoluble, those that are moderately soluble and those that have good solubility in water.

Hydrogen, oxygen and nitrogen are all fairly insoluble in water because they are unable to react chemically with the solvent. Oxygen is carried by haemoglobin in the blood stream and not by solution. If the haemoglobin count falls, the body cannot receive sufficient oxygen from the solution to fulfil its metabolic requirements. Nitrogen can be dissolved in blood under pressure to produce the 'bends', a condition familiar to divers, but at sea level it presents no solution problem to man or livestock. Hydrogen is not a natural constituent of air and can be discounted.

Carbon dioxide and sulphur dioxide will react chemically with water to produce weak acidic solutions of carbonic and sulphurous acids. As the temperature of the solutions is raised the gases are released and under controlled conditions give satisfactory aeration and preservation of a food product.

Ammonia and nitrogen dioxide rapidly dissolve in water to produce strong alkaline and acidic solutions respectively. Ammonia solution of s.g. 0.88 is a very strong source of the gas and should be handled with care if it is used as a detergent liquid. Nitrogen dioxide with its powerful bleaching action was used for flour bleaching but has been largely superseded by chlorine dioxide, which decolorises the bran as well as the endosperm.

COMPRESSIBILITY

The application of pressure to gases causes the molecules to collide, and the frictional forces produced result in heat. When the heat is removed, the gas may liquefy at room temperature; but in most cases low temperatures are needed before the liquid state is reached. Gases may be classified as (a) easily compressible; (b) moderately compressible; and (c) compressed with difficulty.

Nitrogen dioxide can be liquefied at room temperatures if the gas is initially chilled in a freezing mixture of ice and salt. Sulphur dioxide can also be liquefied at room temperature using only 10 atm pressure; it may be retained in a soda-water siphon. Carbon dioxide and ammonia require low temperatures and moderate pressure before liquefaction occurs.

Ammonia has been replaced by 'Freon' for refrigeration purposes as this responds well to liquefaction using less pressure than ammonia and is not toxic.

Oxygen, hydrogen and nitrogen need extremely low temperatures and high pressures to liquefy them. Because of the high pressures involved steel containers are needed to contain them.

The transportation of these gases as liquids is possible. But hydrogen, with its very low boiling point (below $-250°C$), would be difficult to transport over long distances without economic loss.

EXERCISE 9

1. What aspects of texture would you consider to be important in:
   (*a*) tinned pineapple; and
   (*b*) meat balls?
   How would you attempt to evaluate the samples?
2. Discuss any physical differences you might expect between bread made by using
   (*a*) Peerless vertical mixer; and
   (*b*) Tweedy mixer.
3. What effect has a knowledge of the rheological properties of a product upon the quality of its production?
4. Indicate some of the desirable features that should be possessed by a food packaging material. How would you assess these features at a laboratory level?
5. Under what conditions can a knowledge of the plastic and elastic nature of meat or meat products be turned to economic advantage?
6. Compare the advantages and disadvantages of adhesion in a plant bakery or in a meat pie factory.
7. How could the hydrogenation of oils to produce fats suitable for
   (*a*) creaming purposes; and
   (*b*) frying purposes,
   be controlled by viscometry? What differences would you expect between the two types?
8. Distinguish between absolute and relative humidity. Give two applications of the importance of correct humidity in producing good quality bacon and ham.
9. What humidity conditions are likely to prevail in the various manufacturing stages of
   (*a*) milk chocolate; and
   (*b*) fancy cakes?
   How can they be realised in practice?
10. Write brief notes on
    (*a*) diffusion; and
    (*b*) solubility as applied to the gaseous state.
    What practical value has gas diffusion got in the preparation of
    (*a*) sausage rolls; or
    (*b*) cream crackers?

10 Heat kinetics

GAS LIQUEFACTION

The basic principle behind the liquefaction of all gases is the removal of internal energy from the molecules so that eventually they can approach and condense to the liquid condition. When a gas has a high critical temperature it can be liquefied by pressure alone. Chlorine has a critical temperature of 146°C and is readily liquefied by compression to a dark greenish-yellow liquid which boils at −33.6°C. It is commonly sold as a liquid under pressure in steel cylinders, and can be used to sterilise water supplies in the canning industries.

The liquefaction of a 'permanent gas' such as hydrogen or nitrogen is a much more difficult process. A 'permanent gas' is one which cannot be liquefied by pressure alone as it has a very low critical temperature. The figure for nitrogen is −146°C, in contrast to +146°C for chlorine.

Liquid air has as its principal components two important gases, oxygen and nitrogen. In order to liquefy it, both pressure and cooling conditions are needed. Two methods are available: (a) air could be passed through a cold bath containing a more easily liquefied gas, which is boiling at a reduced pressure and has therefore a very low temperature, or (b) by use of the Joule–Kelvin effect. For a variety of technological details of a mechanical nature the second course is preferred.

As the pressure falls on a gas it experiences a cooling effect. Therefore work must have been done in the separation of its molecules. This work is done at the expense of the kinetic energy of the molecules; as a consequence there is a fall in temperature and the molecules come closer together. Although the cooling of a gas in free expansion is small it can become cumulative using the correct practical conditions and so produce a great temperature fall. Figure 27 is a simplified diagram of the process.

The air is first freed from carbon dioxide and water using solid caustic soda. This is necessary otherwise they would condense and block the internal copper coil. The pure air is compressed to about 150 atm, the heat generated being removed by the cooling water, and the air passes on into the copper coil portion which is surrounded by lagging. This part of the internal copper coil is surrounded by an external copper coil to form a double coil. The cooled air emerges from the nozzle which can have its opening adjusted from the outside. The nozzle lies inside a Dewar vessel or thermos flask.

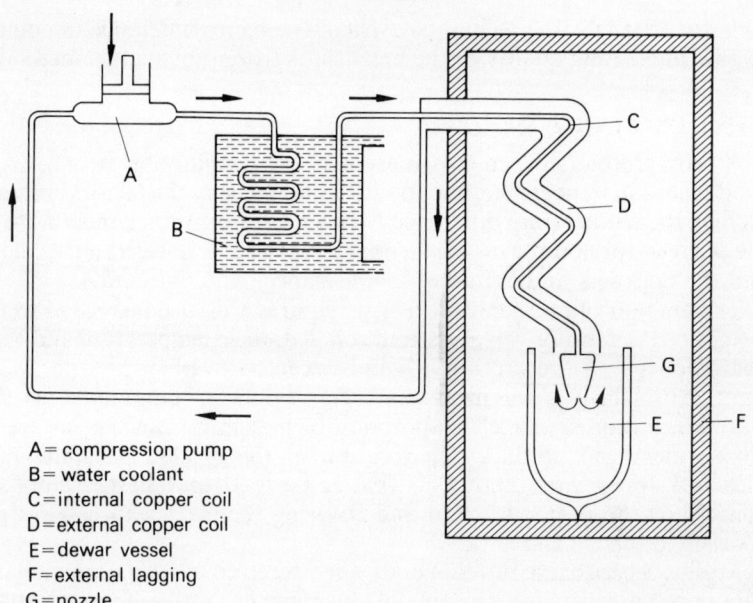

A= compression pump
B= water coolant
C=internal copper coil
D=external copper coil
E=dewar vessel
F=external lagging
G=nozzle

Fig. 27.   Liquefaction of air.

As the air emerges it expands and is cooled by the Joule–Kelvin effect described in the previous paragraph. It then passes upwards along the external coil and, as it does so, cools the incoming gas. The incoming gas is thus cooled before making its expansion, and after its expansion becomes cooler still. After escaping through the nozzle it cools the following gas still further. The cooling of the escaping gas continuously helps the cooling of the arriving gas, and the cooling is therefore regenerative. Eventually the gas emerging from the nozzle cools below the critical temperature, and since the actual pressure, 150 atm, is well above the critical pressure, 39 atm, the gas liquefies and collects in the flask. The heavy external insulation prevents heat coming in from outside.

Liquid nitrogen can be obtained from liquid air by fractionation. It may be used as a cryogenic liquid for the storage of semen for artificial insemination purposes. This selection of semen is essential for producing animals with good beef and dairy characteristics.

In the preservation of chickens, liquid nitrogen falls as a cascade on the samples as they pass on a variable speed conveyor belt which runs through an insulated tunnel. A thin but very protective layer of ice is formed, sealing off the chicken from spoilage organisms.

Thick raw beef cuts (2 cm) can be rapidly frozen using nitrogen spray freezing, the final freezing temperature being of the order of $-30°C$. Weight loss during freezing is much lower using liquid nitrogen than with air blast freezing techniques. There is also no difference in colour, texture and eating quality of the meat slices frozen by the two methods.

HEAT EXCHANGER

A refrigerator's efficiency is measured by its ability to extract, from food placed in it, heat at an economical rate. Some of the factors involved include the temperature difference between the operating temperature of the refrigerator and the external room temperature, refrigerator capacity and the thickness and nature of the insulating material used.

The amount of refrigeration energy required will also be related to the weight of the food or carcasses to be chilled, their temperature and their heat capacity. A freezer unit should never be overloaded with warm meat, particularly if the meat has been ground or comminuted as this will induce food spoilage. Evaporation of moisture from the surface of frozen meat can produce deterioration of the product resulting in a bleached appearance, known as freezer burn. It may be prevented by controlling the heat extraction and covering the meat with a skin-tight covering or an ice glaze.

Milk is a perishable product and, when received at a processing plant, should be cooled as soon as possible by passing it through some form of heat exchanger. Usually this consists of a large number of plate heat exchanger units assembled on a filter system. Each plate has a large surface area of corrugation. Milk enters at the top and flows over the corrugations, cool milk leaving at the lower portion. A combination of large numbers of plates with powerful turbulence and counter-current flow rapidly chills the milk. The plates are electrolytically treated to give a good hygienic surface. Heat exchanger plates are also used by the fruit juice and brewing industries.

Fresh water has to be used as the coolant because sea water has a corrosive effect on the plates. Titanium is used in the plate construction, and although it is expensive it is very durable compared with cast iron plates.

SUBLIMATION

Solid carbon dioxide at normal temperature is an ideal refrigerant for the ice-cream industry. A small block placed in the base of an insulated container will keep packed ice-cream cartons in first class condition. Unlike ice, which produces water on melting, solid carbon dioxide at normal pressure passes direct into the vapour condition without going through the intermediate liquid state. This process is known as *sublimation*.

Meat, fish, fruit and vegetables can be dehydrated rapidly under vacuum conditions. The product, however, may not reconstitute well and therefore has a poor acceptability rating. If the product is frozen initially and then transferred to a vacuum chamber and heated slightly, the ice will sublimate. No moisture is produced to harbour food spoilage organisms and in addition the dehydrated food is stable so long as air and moisture are kept from it.

The process of 'accelerated freeze-drying' (A.F.D.) requires both technical skill in its operation, and expensive equipment. However on a large turnover the improvement in quality and subsequent sales appeal will produce a satisfactory economic return. Commercial equipment is available in which heat is supplied by a plate heat exchanger through which warm water is circulated.

It is essential that meat be presented in thin portions if the texture is close. Pork may be frozen after being boned out and treated successfully by A.F.D. *providing* the loins are cut into chops about 15 mm thick on a bacon slicer. This appears to be about the correct size, thinner slices tend to go brittle. Legs and fillets of lamb are too thick to be successful and are best held in the cold store at about −20°C.

REFRIGERATION

It is a common fallacy that refrigeration will preserve food indefinitely. Refrigeration and deep freezing will not kill bacteria. Their metabolic activity is much reduced but only over a short period of time, and unless very low temperatures are used, decay will result. The normal refrigeration temperature range for most domestic appliances is −5–+5°C with a small freeze compartment which can fall to about −10°C.

Unpleasant 'off-flavours' will make their presence felt if food is not consumed on a rota basis. If the temperature of meat is reduced below −2°C, it will freeze. This produces a change in the physical state of the tissue and also affects the rate of enzymatic and chemical changes. Chilled beef approximately −1.5°C contains appreciable amounts of free water. In order to preserve it, a deep freeze is needed. In a deep freezer temperatures of −20°C or below are a common feature. At this level the food will keep for several weeks, indeed months, depending upon the conditions pertaining prior to refrigeration.

All refrigerators contain a closed coil system in which a suitable refrigerant can be contained. Figure 28 is basic to all types, though in place of the liquid coolant in the condenser coil section a fan may be found which removes to the external atmosphere heat produced by compression.

Refrigerants in commercial use include ammonia, methyl chloride, carbon dioxide and various fluorinated hydrocarbons known as 'Freon'. Ammonia is being replaced by Freon by many firms, the latter being non-toxic and safe with all kinds of food.

The refrigerant is compressed to produce a cold liquid and gives up its latent heat to the liquid coolant which surrounds the condenser. This high-pressure liquid then enters the expansion valve where pressure is allowed to fall. Low pressure liquid enters the evaporator coil where it

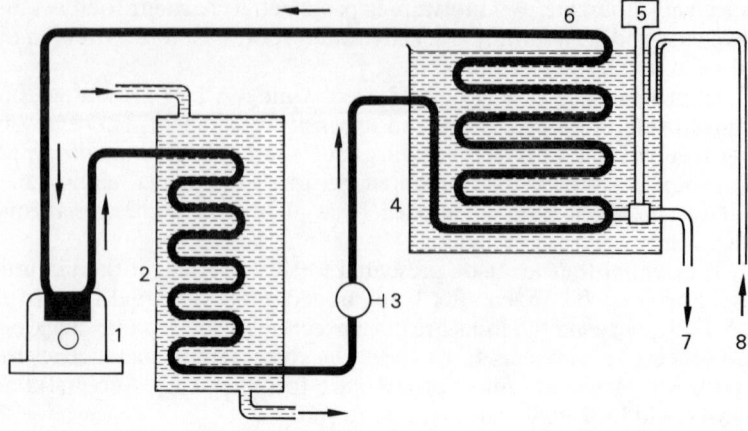

1. pump
2. condenser coil and coolant
3. expansion valve
4. brine tank
5. brine circulator
6. evaporator coil and vapour return
7. freezing brine outlet to cold store
8. warm brine return pipe

Fig. 28.   Refrigeration cycle.

begins to evaporate, drawing the necessary latent heat of vaporisation from the brine which, in consequence, falls to well below 0°C. Cold brine is then pumped to the cold store which contains the goods on shelves. Heat is absorbed from them, and warm brine returns to the evaporator coil section. Refrigerant vapour then returns to the pump and is ready to begin its cycle again as liquid.

Other refrigerators may use a gas flame or electric heat in place of a pump to provide the energy needed to circulate liquid and vapour. However the result is the same, cooling by vaporisation under reduced pressure.

It must be emphasised that refrigerated counters are not storage units. Their purpose is to maintain the product in prime condition and freshness while on display. Counters and cabinets should be cleaned daily and their temperature checked at various levels inside the display area.

There is a tendency to overload the display; on no account should food be placed above the load lines. Certain products, for example bacon, are enhanced by the presence of permitted coloured lights. Refrigeration temperatures are therefore critical since a power failure would not reveal itself in colour changes on the product if the lighting system was on an independent circuit. Every product will have an ambient temperature which has been found suitable in practice. In the case of bacon 5°C would be appropriate for a daily display only whilst −4°C would maintain good quality for approximately 2–4 weeks.

Cold storage warehouses are in reality little more than large refrigerated rooms. Good insulation is needed. Cork is still used but because of its porous nature and susceptibility to decay it is being replaced by synthetic materials such as plastic sheeting and fibre glass. It is important to prevent heat entering from the outside, and to ensure the removal of internal heat. The maximum working temperature is of the order of −18°C (0°F), that is for quality purposes the temperature should not go above this level.

Quality aspects of frozen blocks of cod fillets at storage temperatures of −18, −24, −30 and −50°C for times of up to 4 years have been observed. In all cases quality decrease had occurred but at the lowest temperature and even after 4 years the sample could have been eaten. There is a tendency for tallowing flavours to be produced which would prohibit sales to the general public, yet it is a remarkable instance of the success that can be achieved with a very perishable product.

Vegetables lend themselves very successfully to the technique of quick freezing. If vegetables are frozen slowly, particularly between the temperatures of 0°C and −5°C, their water content produces large ice crystals on freezing. As ice formation results in a 10 per cent expansion the product will, on thawing, contain large areas of texture damage. A combination of oxidation and spoilage organisms can rapidly produce deterioration in quality. However, if food is frozen rapidly so that it passes with the minimum of time through this zone of maximum ice-crystal formation then texture damage will be at a minimum, with optimum colour and flavour potential maintained. Changes in texture are also related to the osmotic shrinkage of cells during freezing.

With peas the crop is harvested when a favourable result is achieved using a tenderometer. A tenderometer consists of a number of blunt steel blades which are pressed against the peas. The pressure needed to force the blades through is noted and is a measure of the tenderness of the sample. When conditions are favourable the crop is quickly harvested and within less than 2 hours is ready for quick freezing.

Although bacterial activity is arrested at the temperatures used in quick freezing, a large number of enzymes are still active. Peas are subjected to a bleaching process by scalding them for a short time in boiling water, a process which destroys most of these troublesome enzymes.

Tests have been conducted to show how blanching affects the quality of the pea when subjected to a quick-freezing treatment. The five most common pea varieties were examined for suitability for quick freezing and their keeping times at various temperatures were measured.

The peas were blanched at 95°C for 2 minutes, packed in cardboard boxes and quick frozen at −30°C. Storage temperatures were −30, −20 and −10°C. Colour was measured immediately after picking, by spectrophotometry of the acetone extract. In each variety, compared to raw peas, blanching caused a drop of about 50 per cent in consistency. A comparison of the texture of quick-frozen green peas to the blanched peas prior to freezing showed a hardening of the texture during storage.

These results combined with organoleptic taste tests, and on the deterioration rate as a function of storage temperature, enable a decision to be taken as to the need to blanch or not. In the tests discussed it was found that a storage temperature of $-20°C$ was adequate. This is an important economic factor since the maintenance of very low temperatures is costly. If food can be preserved adequately at $-20°C$, bearing in mind the supply and demand factor, it is foolish to hold it at $-30°C$.

Modern quick-freezing plants usually rely chiefly upon two methods for maximum efficiency, namely the tunnel freezer and the multi-plate freezer.

In the tunnel freezer very cold air blows up through the food which passes along the tunnel on a mesh belt. By having the air injected upwards through the food small particles are kept in continuous motion and adhesion prevented. If the food is placed on a fluidised bed which produces a vibratory action, small items like peas behave like ping-pong balls bouncing on a table. Tunnel freezers are often employed for foods which need only a short freezing time. Sliced beans would be processed satisfactorily in about 10 minutes. By comparison, pies and meat joints may be successfully treated using freezing times ranging from 1 to 3 hours.

The multi-plate freezer has a wider range. It is a cabinet with a number of hollow shelves which form a freezing surface. It is through these hollow plates that the refrigerant is circulated. Usually the product is loaded on to the shelves in packaged form and by utilising hydraulic pressure the plates are brought in close contact with the product and rapid cooling results. The nature and thickness of the product to be frozen affect the time; meat and fish may take upwards of 45 minutes.

Freezers are also available which use liquid carbon dioxide on a continuous spiral belt conveyor system. Increased capacity is obtained from the spiral effect and fast freezing takes place at about $-63°C$. There appears to be less than 1 per cent moisture loss which compares favourably with conventional air-blast systems where dehydration losses are frequently higher. It is claimed that this will produce juicier, better flavoured and red-coloured meat and also fresher tasting poultry and seafood. Certainly the time taken to reach the operating temperature of $-63°C$ will be much less than that for blast freezing.

Liquid nitrogen has been used to freeze food by direct immersion, particularly in the U.S.A. It is expensive but may be used on foods which do not freeze entirely satisfactory by other methods. Strawberries and tomatoes, because of their delicate nature, respond well to immersion. However, it must be noted that, unless the fruit is used quickly, the superior flavour will rapidly fade and after three months storage the quality is no better than that offered by more conventional freezing techniques.

Refrigeration in its many forms has played, and will continue to play, an ever increasing role in food preparation on a world-wide basis. It enables food to be harvested in plenty and stored until either leaner times arrive or else preserved in bulk for a national emergency. Admittedly the flavour and general quality of frozen food is not quite as good as fresh food but this is more than compensated for by the wide variety of products that can be eaten out of season. Frozen food is also more expensive than the natural equivalent since larger amounts of electrical energy are involved.

EXERCISE 10

1. What impact has been made by gas liquefaction upon the food industry with respect to meat quality?
2. Discuss some physical principles important in the construction of heat exchangers. What virtues might be possessed by graphite as a material for a heat exchanger?
3. How does A.F.D. compare with other conventional methods for dehydrating food? What methods would be suitable for the preparation of
   (a) milk powder; and
   (b) condensed milk?
4. Write an account of any two *different* refrigerated techniques you have seen used in a food factory. Give reasons why you think the manufacturers selected their particular method.

11 Light mechanics

WAVELENGTH AND COLOUR

The concept of wavelength can be visualised by reference to a piece of string which is fastened at one end to a post, the free end being held in the hand. By moving the hand up and down the illusion of forward motion is created; the string appears to travel in a wave motion. The type of wave produced is a *transverse wave* because the vibration is at right angles to the direction of motion.

On examination the string appears to consist of a series of crests and troughs which travel forwards at a given velocity, called the *wave velocity*. If the string is imagined to be composed of hundreds of different particles the distance between any two particles on two separate crests or troughs constitutes one wavelength (λ). On our string analogy this distance would be relatively easy to measure, but the wavelengths encountered with light waves need a much finer unit—the Ångström. It has a magnitude of 10^{-8} cm and is often used to express the size of molecules.

The faster the string is vibrated or oscillated the higher is its *frequency*. The number of complete oscillations made in one second is called the frequency (f), the SI unit of frequency being the hertz (Hz), defined as 1 cycle (or oscillation) per second.

All the waves encountered in food preparation or preservation have their origin in electric and magnetic vibrations; hence they are known as *electromagnetic* waves. A useful formula for the calculation of either the frequency or the wavelength of a light wave is:

$$v = f\lambda,$$

where v = velocity of light at 3×10^8 m/s.

Example

Calculate the frequency of microwaves of wavelength 12 cm.

$$v = f\lambda$$
$$\therefore f = \frac{v}{\lambda} \quad \frac{3 \times 10^8}{0.12} = 2.5 \times 10^9 \text{ Hz.}$$

This may now be expressed in a larger frequency unit, the megahertz (MHz) where 1 000 000 Hz = 1 MHz:

$$\therefore 2.5 \times 10^9 \text{ Hz} = 2500 \text{ MHz.}$$

142

Microwave ovens are a feature of many first class restaurants. They can produce in a matter of minutes a hot meal which has been pre-cooked and preserved in deep freeze. The food absorbs energy from the oven by a kind of molecular friction. One limiting factor with this type of heat is the constitution of the food involved. Sponge pudding responds well as it is homogeneous but a meat course involving lean, fat and gristle would present problems because of its uneven constitution.

Figure 29 illustrates a section of a wave motion and also the tremendous wavelength span of the electromagnetic spectrum.

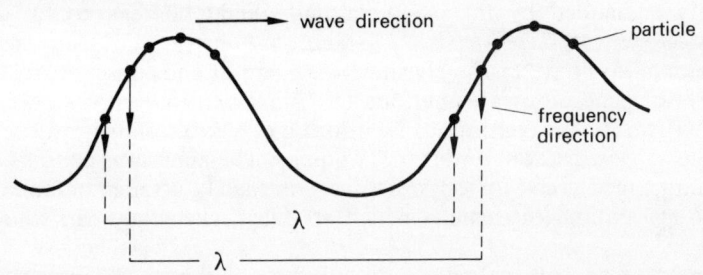

λ = wavelength between two particles of identical height

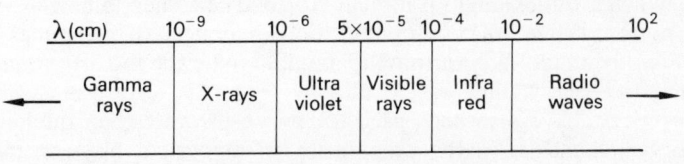

Wavelength range (not to scale)

Fig. 29. Electromagnetic waves.

The sensation of colour is the result of electromagnetic radiation of a particular wavelength falling upon the eye and being interpreted by the optic nerve. It is a very small range of the wavelengths available and covers approximately 4000–7500 Ångström units, with violet at the lower limit.

Colour represents an important feature in man's appreciation of food, which to be fully enjoyed must not only taste good but look good. When food is in short supply fat is withdrawn from the animal's body but the orange carotene pigment remains and as a consequence the colour of the fat darkens.

Muscle colour is a variant factor; those muscles which are used for prolonged period of activity, such as the psoas muscle, are darker in colour than others which are less used. Muscles found in old animals and the horse are dark and the texture is correspondingly tough. A pale-coloured muscle will produce mild flavour and have the correct degree of 'chewability'. Larger and older animals always give tougher meat than younger beasts. One may therefore regard colour as a visual factor of food quality.

Occasionally, when cured or cooked meat is cut, there may appear a greenish iridescence on the surface. This is caused by light refraction at the thin layer of covering fat and does not indicate spoilage. Bacterial growth on cured meats can produce a non-iridescent green colour. Although in both cases the meat is usually safe to eat, it is not of the quality demanded by the customer and would be rejected for sales purposes.

Gamma rays have extremely short wavelengths and consequently high powers of penetration through matter. A recent discovery has been to utilise them for the continuous monitoring of a series of articles, such as cartons of powders or containers of liquids. The apparatus consists of a constant speed conveyor on which are alternately arranged aluminium block spacers and the articles which are filled when they pass under a filler.

A gamma-ray source directs radiation through the spacers and articles. The signal produced is then compared with the signal expected from a unit of average weight. An instantaneous comparison is made and any article which is underweight is directed to an article rejection mechanism. There is no adverse effect upon the food or danger to personnel. The system can be made fully automatic and eliminate the monotonous task of a human check-weigher.

Microwaves have also been used for purposes other than the heating of pre-cooked meals. In the preparation of arrowroot biscuits, microwaves will reduce their moisture content to about 2 per cent in just over 4 minutes and produce the crispness characteristic of this type of biscuit. Microwaves are also used to detect the water content of butter. In this instance the butter acts as a resonant structure for microwaves, and the amount of water in the butter affects the frequency of the waves used. The signal obtained can be interpreted on an oscilloscope for maximum and minimum tolerable amounts of water.

Research has been conducted on the possible commercial use of electromagnetic energy in the control of insects. At the present time there is considerable concern over the environmental contamination caused by pesticide residues. Manufacturers of food products need to exercise particular care to keep such chemical agents out of their finished products and they must also guard against insect fragments.

The milling and baking industries are those chiefly concerned with insects that can contaminate starch products. At the high energy end of the spectrum, X-rays have been found useful in detecting hidden insect infestations in grain and exercise a sterilising effect on the population. However, since X-ray production is expensive, practical use for insect control does not appear to be a possibility. Gamma radiation which produces the same effect, has been more interesting, because it is abundantly available from certain radioisotopes. Gamma sources, while still expensive, can be produced to handle irradiation of large quantities of material.

Considerable work has been done with ionising radiations to produce sterile male insects in controlled conditions and to increase the level of energy to a lethal dose for most species of stored-grain insects. Success was highest in dried fruit, without any damage to the fruit. However, this is not at present a viable economic proposition.

Ultraviolet radiation, often called blacklight, has been used to detect contamination in cereal products because rodent excreta and some other contaminants fluoresce under ultra-violet radiation. Principal interest in using visible and ultra-violet radiation for insect-control applications has depended on the use of such radiant energy to attract certain insect species. The use of radiant energy fly traps in bakeries ensures control of any insects which may enter the works during the intake of raw materials. A wavelength most attractive for some insects is about 360 nm (3600 Ångström units) in the near ultra-violet. Experiments with stored-grain insects showed the Indian-meal moth responds to this wavelength. Other species were attracted to a visible wavelength of 500 nm.

Infra-red radiation from electric infra-red lamps can effectively control stored-grain insects. Infested material can be passed under an infra-red source on a conveyor with a speed set to provide the correct exposure. For wheat and rice, exposures which produce grain temperatures in the 65–70°C range are necessary to obtain control of all developmental stages of stored-grain insects. Again cost is a limiting factor to extensive application.

There has also been some speculation that insects utilise infra-red radiation in their communication processes. If this should be verified, new approaches for insect control might be considered. At present such possibilities appear unlikely for stored-product insects.

The low energy end of the spectrum is occupied by the radio waves. Radiofrequency (R.F.) energy has been tested for its insect-control capacity over a wide range of frequencies. Although success has been achieved with insects, without grain damage, the cost is several times greater than chemical control.

One may conclude that radiofrequency, infra-red and gamma-radiation regions are all potential control methods for insect population. All three types of radiation appear to be more costly to use than currently employed methods. They do, however, possess the advantage of achieving control of existing insect populations without damaging the host product or leaving undesirable residues.

The potential for application of electromagnetic energy to limit insect population hinges on economic factors and continued acceptability of current practical methods. Continuing research on the interaction of electromagnetic radiation with biological systems may lower the cost and uncover new mechanisms which may be vital in controlling insect damage to food estimated to cost millions of pounds annually.

POLARIMETER

As we have previously observed, certain compounds are able to affect the direction of a beam of polarised light. Optical isomerism is possible because of the presence of asymmetry in such compounds.

The polarimeter or polariscope is an apparatus specifically designed to measure the rotation of the plane of vibration if polarised by any optically active substance. Monochromatic light is used in observations; this is light consisting of vibrations of the same, or nearly the same, frequency and is light of one colour. Usually the source of mono-chromatic light is a sodium lamp which emits a strong yellow beam at a wavelength of about 5890 Ångström units. The light enters the instrument and is made parallel by a lens. It then passes through a fixed Nicol prism which polarises it. The radiation then passes through the optically active solution, contained in a glass tube of known length, and through a second Nicol prism (the analyser) which can be rotated about its axis. Visual observation of the path of light is completed by using a series of lenses.

Initially the two Nicol prisms would be in the crossed position (total extinction) without the sample in the polarised beam. The sample is now introduced and the analyser is turned until the prisms are again crossed. By reference to an angular scale, attached to the rotating analyser, the angle through which the analyser has passed is observed. This rotation will be directly proportional to the length of column through which the light passes, the temperature and wavelength used and also concentration. The specific rotating power of the substance will therefore be unique to a given set of conditions.

SACCHARIMETER

The polarimeter could be modified to give useful data on optically active nutrients of the carbohydrate and protein types but was not specifically made for this purpose.

Sugar is a major raw material in confectionery work and the quality of finished goods may be directly related to the type and amount of sugar used. Indeed starches can be included which with the sugars embrace the realm of carbohydrates. The optical activity of each carbohydrate will be a function of the variants discussed in the last paragraph. A *saccharimeter* (not to be confused with a saccharometer) was devised as a specialised polarimeter.

The saccharimeter is commonly employed for sugar analysis. An essential difference between the polarimeter and the saccharimeter is that the polarimeter employs monochromatic light and is read in angular degrees, whereas the saccharimeter uses white light and, in addition to the angular reading, has a percentage sugar scale for concentration purposes. The concentration scale is usually calibrated in international sugar units. For this purpose a normal solution of sugar is prepared. This is a solution of 26 g of sucrose in a volume of 100 cm^3. When this solution is placed in a tube of 200 mm length and examined at 20°C it will give a sugar reading of 100. Therefore a half normal solution will give a reading of 50. This sugar scale is used directly to measure sucrose concentrations, or indirectly instead of angular degrees to interpret other sugar readings.

Every sugar requires a different amount dissolved in water to give a reading of 100 which by definition will be its normality by optical observation. For glucose the normal weight is 32.366 g whilst for maltose it is only 12.515 g. When a single sugar solution is examined readings are direct and simple but this is not the case with food mixtures. The following procedure illustrates how the saccharimeter would be used to find the concentration of a simple syrup for sheen purposes on confectionery goods.

Locate the two eyepieces at the head of the instrument. The top eyepiece can be focused to show the scale and index line. Look through this eyepiece and turn the control wheel until the scale reads zero on the upper scale. Without disturbing the setting, transfer the eye to the lower field eyepiece and focus to obtain a circular disc of light. By means of the control wheel, move the scale a few degrees about the first position, observing the effect on the field. It will be observed (a) that the field is divided into two halves rapidly changing in intensity as the control is moved, and (b) that between the extremes of black-out of the two halves, a position will be found that gives equal intensity. This is the balance position and is used in all measurements. Readings are seen more clearly in the dark.

The instrument is now set up on the bench with the angular head facing the operator and the illuminated sodium lamp in direct line with the axis of the instrument and about 10 cm from the end.

A portion of the syrup is now placed carefully in the saccharimeter tube, making sure that no air bubbles are trapped. Look through the field eyepiece and refocus if necessary. The control wheel is now adjusted to match exactly the two halves of the field, and the angular scale reading and sugar reading are observed through the top eyepiece. A simple calculation can now convert the reading to a percentage concentration.

SPECTROMETER

The identification of radiation from a substance within the visible wavelength range is possible using a spectroscope and the spectra obtained can be analysed. A *spectrometer* is a type of spectroscope which is calibrated to analyse the light making use of the different refractive indices of its components. It can be used for the examination of the spectra of hot gases and other sources of light.

The instrument is made from heavy metal so that it will remain steady during the analysis procedure for spectra. There are three main parts, collimator, table and telescope (see Fig. 30).

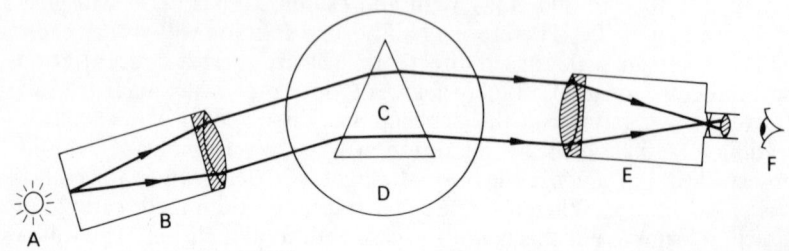

A = radiation source
B = collimator
C = prism
D = table
E = telescope
F = observer

Fig. 30. Spectrometer.

Radiation from the object under test enters the collimator through a narrow slit, the width of which can be adjusted by an external screw. A convex lens is placed in the collimator section so that the slit will be at its principal focus; light passing through the lens will thus be in parallel beams. (The verb collimate means to make parallel.) As light enters the prism, because each colour has a different refractive index, separation of colours occurs into different parallel beams, each of which is brought into its own focus by the second lens in the telescope section. Each individual colour is seen as a vertical band which is really an image of the slit and is viewed through the magnifying eyepiece. The observer will therefore be able to identify different substances by an analysis of their emission spectra.

Before the spectrometer can be used three adjustments are necessary. First, the collimator must be adjusted so that parallel light emerges from it. Secondly, the telescope section must be adjusted so that parallel rays entering it are brought to a focus on the cross-wires near the eyepiece. Thirdly, the refracting edge of the prism must be parallel to the axis of rotation of the telescope. This is done by levelling the table on which the prism is mounted, using moveable screws attached to the table base. The collimator is fixed, but the table and the telescope can be rotated around a circular scale graduated in degrees.

As a physical method of analysis of mineral food additives or metal contaminants the spectrometer gives qualitative information about these substances.

In the analysis of food mixtures, for reasons previously discussed, the more sophisticated spectrophotometer is used. This will produce a quantitative run-out and gives more reproducible data.

EXERCISE 11

1. Explain the term electromagnetic wave. How can such waves be used to assess the quality of meat?
2. What advantages and disadvantages are likely in the use of microwaves for food preparation?
3. Distinguish between infra-red and ultra-violet rays. How could they be of use in flour confectionery?
4. What problems are posed by the use of very high frequency electromagnetic waves in controlling insects found in bulk storage of raw materials?
5. How would a saccharimeter be of value in the quality control of
   (*a*) fondant; and
   (*b*) syrup in canned fruit?
6. What physical features are important in the construction of a spectrometer? Give two limitations in its analysis of a food mixture.
7. How is a working knowledge of optics useful in the quality control of
   (*a*) chocolate;
   (*b*) cake mixes; and
   (*c*) fats?

12 Magnetic and electrical topics

MAGNETISM PRODUCING ELECTRICITY

When electricity passes down a conductor a magnetic field surrounds the conductor carrying the current. A compass needle will be deflected if placed in the vicinity of the force field. Conversely, the creation of a magnetic field round a closed circuit will cause a current to traverse the circuit.

Michael Faraday in 1831 was the first person to demonstrate practically the electro-magnetic relationship suggested in an earlier work on magnets by Oersted in 1819. A modern adaptation of Faraday's experiment would involve taking a long length of cotton-insulated or plastic-insulated copper wire and making a coil of some 200 turns. If the two bared ends of the wire are joined to a sensitive galvanometer, and a strong bar magnet is swiftly plunged into the coil, a small deflexion of the galvanometer needle is observed. Electricity has been created by the motion of the magnet, but no electricity flows when the magnet is stationary inside the coil. When the magnet is swiftly removed the galvanometer needle is deflected in the opposite direction.

Faraday also observed that the faster the magnet is moved the greater is the deflexion, though it persists for a shorter time. Finally, he noted that reversing the magnet, that is, plunging the opposite pole into the coil, has the same effect as reversing the direction of motion; it produced an opposite galvanometer deflexion.

ELECTROMAGNETIC INDUCTION

Faraday's experiments showed that electricity was produced by a moving magnet and because the magnet was not physically in contact with the insulated wire the current was said to be *induced*. This new technique of electromagnetic induction in its modern form is the basis of every modern electrical appliance that can be found in the food industry from food mixers to closed-circuit television and computer banks.

Further work by Faraday showed that induction can take place without the use of a magnet. Figure 31 outlines an experiment which reproduces his findings.

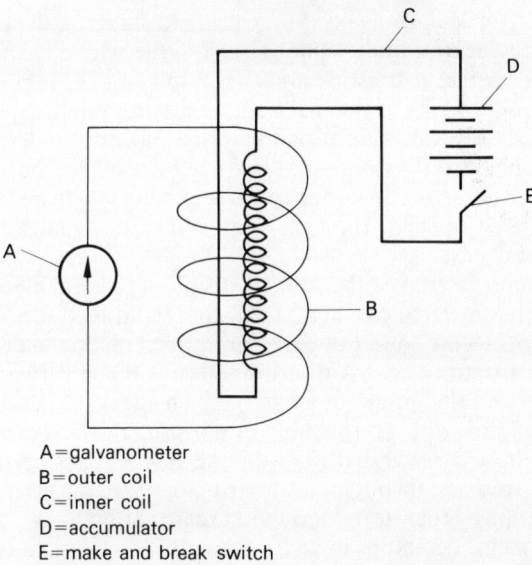

A=galvanometer
B=outer coil
C=inner coil
D=accumulator
E=make and break switch

Fig. 31. Electromagnetic induction.

A long thin wire which is coiled in the middle has its ends attached to a source of electrical power (three accumulators). This coil is inside a second coil which has fewer but larger turns at its middle portion and also has a very sensitive galvanometer in its circuit. When the current through the inner coil is switched on, and the magnetic field associated with it grows, we observe a momentary deflexion in the galvanometer connected to the other coil. When it is switched off, an opposite deflexion is obtained. The results are therefore identical with those obtained using a moving magnet and represent a more practical way of producing electricity by induction.

The experiment can be repeated with a piece of soft iron placed inside the inner coil. Enormously increased deflexions are obtained and a less sensitive galvanometer must be used to avoid damage to the instrument. Quite obviously, the magnetism induced in the iron increases the effect.

Faraday's experiments show that an e.m.f. (electromotive force) is induced in a circuit when the magnetic field passing through the circuit changes and that the greater the rate of change, the greater the e.m.f. Today we refer to Faraday's Law of Electromagnetic Induction which states: when the magnetic field through a circuit changes, an e.m.f. is induced into the circuit which is proportional to the rate of change of the field.

An ingenious application of induction has recently been developed for checking the pressure of carbonated beverages in cans inside a sealed case. This test for leakage is necessary prior to shipping the product. It is designed to replace an audio-manual technique whereby the top of the can is 'tapped' with a drumstick and the sound emitted is compared with the sound produced by a normal can. Since the pitch depends on the pressure within the can, variations in pitch indicate undesirable variations in pressure and thus potential leakage or spoilage, but this method is open to human error. In order to inspect cans inside a shipping case, the case must be opened, the cans removed, tested and the good cans placed in a new case.

The electronic version of the test substitutes a pulse of electromagnetic energy for the mechanical action of the hand-held drumstick. The operator listens to the sound of each can tapped in a manner similar to a doctor using a stethoscope. A discrimination is made between the 'ping' of a good can and the 'pong' of a bad one. The device works through the top of a sealed case without touching or harming the beverage in any way so that only those cases actually containing suspect cans need be opened. It is claimed to work through cardboard, corrugated fibreboard, plastic shrink film or any other non-conducting material.

Electrical metal detectors have been used in confectionery lines for a number of years. As each package passes under the electronic eye (or scanner device) the merest trace of ferrous metal will affect the signal and cause a plate under the sample to open and reject the sample into a disposable bin. Tests at present would indicate that a metal detector is capable of being developed which will recognise all metallic impurities, whether they are metallic or not, provided the material being scanned is non-metallic.

RESISTANCES

Materials in the food industry will offer some resistance to the passage of an electric current. The extent to which a conductor resists the flow of a given current depends upon its physical dimensions, the nature of the material of which it is made and its temperature. Normally resistance will increase as the temperature rises. This information is used to produce a convenient and accurate thermometer, in which the temperature is deduced from the measurements of the resistance of a spiral of a metal (usually platinum) in the form of a wire.

A piece of wire of fairly high resistance is often used in a circuit to limit the current to a suitable value. Such a piece of wire is called a *resistor*. It is often convenient to use a resistor which can be adjusted to have a suitable value of resistance, such an instrument being called a *rheostat*. Electrical hoists or trucks need to have their power controlled by the installation of rheostats.

The ohm is the SI unit of resistance and is the resistance between two points of a conductor when a constant difference of potential of 1 V applied between these points produces a current of 1 A in the conductor.

The wire used in a rheostat is usually nichrome, an alloy of nickel and chromium, which has a high resistance and which is not damaged by high temperatures.

Resistances can be wired in two ways. If the resistors are joined end to end to produce a continuous circuit they are in *series* and the same current flows through each. To calculate the combined resistance the following formula is applied:

$$R = R_1 + R_2 + R_3, \quad \text{etc.}$$

where R is the total resistance and $R_1 \ldots R_3$, etc. are the individual resistances.

When the resistors are placed side by side with their corresponding ends joined together they are said to be in *parallel*. Although they will all receive the same voltage they will only share the main current in the circuit. The reciprocal of the combined resistance is now the sum of the reciprocals of each resistance.

$$\frac{1}{R} = \frac{1}{R_1} + \frac{1}{R_2} + \frac{1}{\mathbf{R}_3}, \quad \text{etc.}$$

Example

Compare the values obtained for two resistors of 3 and 7 Ω respectively when wired (*a*) in series and (*b*) in parallel.

In series
$$R = R_1 + R_2$$
$$R = 3 + 7 = 10 \, \Omega$$

In parallel
$$\frac{1}{R} = \frac{1}{R_1} \quad \frac{1}{R_2}$$

$$\frac{1}{R} = \frac{1}{3} + \frac{1}{7} = \frac{10}{21}$$

$\therefore$ $10R = 21$ or $R = 2.1 \, \Omega$

It is useful to remember that for two resistors in parallel:

$$\text{Combined resistance} = \frac{\text{product of resistances}}{\text{sum of resistances}}.$$

An accumulator needs to supply a very heavy current (about 200 A) to turn the starter in bread or meat delivery vans. A car battery needs to have a very low resistance in order to allow such a heavy current to pass. Thick copper wire has less resistance than thin wire. In the wiring of electrical appliances allowance has to be made for the resistance offered by the wire if a fire hazard is to be avoided.

The use of decorative colour lighting in shop windows necessitates wiring in parallel. In parallel if one light burns out the remainder will still function because of the shared nature of the wiring; in series the malfunction of any individual will stop the current flowing to the remainder.

Ice has more resistance than water. This factor could be used to assess the quality of frozen meat by its conductive capacity.

HEATING EFFECT OF ELECTRICITY

The amount of heat produced when an electric current passes through a wire was investigated by Joule. He showed that the heat produced when a current passes through a wire is proportional to the square of the current, the resistance of the wire and the time.

The heat produced equals the product of the current squared, wire resistance and time. Because heat is a form of work it follows:

$$\text{Work done (or heat produced)} = A^2Rt,$$

where A = current in amperes, R = resistance in ohms and t = time in seconds.

Because 4.2 J equals approximately 1 calorie it would be possible to calculate the heat needed from any electrical appliance to cook various food items, and equate it with the electrical power supplied. The SI unit of power is the watt, which is defined as the rate of working of 1 J/second. Therefore:

$$\text{Power in watts} = \text{rate of working in J/seconds.}$$

Example

A bacon slicer operates on 12 A and has a resistance of 20 Ω. It is on for 30 minutes. Calculate the amount of work done and also the power consumed.

$$\text{Work} = A^2Rt \text{ J}$$
$$= 12^2 \times 20 \times 1800 = 5\,184\,000 \text{ J}$$
$$\text{Power in W} = \frac{\text{work in J}}{\text{time in seconds}}$$
$$= \frac{5\,184\,000}{1800} = 2880 \text{ W.}$$

This calculation is useful if the voltage is unknown. However, when the voltage is available, a simpler calculation would give the power used, since $W = V \times A$. In the preceding calculation the voltage would be 240.

RING MAINS

The old-fashioned and rather expensive method of wiring power points was to use a distributive fuse-box, rather like that used on the lighting side. Separate wires were then run to each socket from an individual 20 A fuse in the distribution box. When an individual appliance failed it could be costly in terms of time wasted before the specific fault was traced; there was also the fire risk involved if a short circuit occurred.

Power circuits always have three wires, the live and neutral mains and an independent earth connection. The modern method of wiring power points is to use a ring main. The wiring from the meter passes to a main switch with a heavy amperage fuse in the live lead. It then makes a complete circuit of the premises eventually arriving back at the mains switch. All outlet sockets are of one size and are connected to tappings from the ring main. Protective fuses in cartridge form are carried inside the plugs used to connect the various appliances to the sockets. Each fuse can then be made to suit the particular appliance. Fluorescent lighting will only need a low amperage fuse compared to a heavy fuse load for a power appliance such as a dough mixer or electric oven.

GALVANOMETERS

Galvanometers are sensitive current-measuring devices which detect, compare or measure small electric currents by utilising the magnetic effect produced when electricity flows.

A simple galvanometer could be made by placing a pivoted magnetic needle at the centre of a number of turns of insulated wire. When an electric current is passed through the coil the magnetic effect produced causes the needle to turn, the turning being related to the strength being passed. This instrument has no practical value because it could be affected by electric motors or power cables in the near vicinity, and it has to be set up in one particular direction because of the controlling influence of the earth's field.

The modern galvanometer is of the moving-coil type (Fig. 32); by using a mirror it can operate as a spot galvanometer for use in detecting the colour grade of flours or for delicate temperature measurements of bacterial growth.

In the ordinary portable type of moving-coil galvanometer the coil is carried on two pointed steel pivots which rotate in cup-shaped sapphire bearings. The friction in this type of bearing is very small and the bearing is also very resistant to wear. Two spiral springs, like the hair-spring of a watch, are fastened at the top and bottom of the coil and provide both the resistance against which the coil turns and also electrical connections to the coil. The springs are wound in opposite directions so that should a change in temperature alter their properties the opposing effects in the two springs will cancel out.

A very light pointer is attached to the coil, and counterweights are provided to balance the weight of the pointer. The magnetic field is provided by a horseshoe magnet, the pole pieces of which are hollowed out into a cylindrical shape. A soft-iron cylinder is supported symmetri-

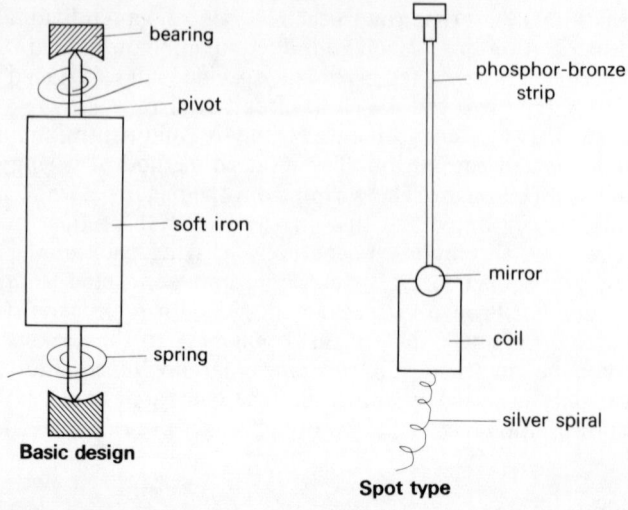

Fig. 32. Moving coil Galvanometer.

cally between the pole pieces by a non-magnetic bracket, and the coil rotates in the ring-shaped gap. Any electrical current received by the galvanometer will magnetise the soft iron and produce rotation between the poles of the magnet; the pointer will then move and current will be indicated on a fine scale often calibrated in microamperes (μA).

When still greater sensitivity is needed, the coil can be suspended by a long thin strip of phosphor-bronze, which also forms one electrical connection, the other being a loose spiral of silver wire beneath the coil. The force needed to twist the suspending strip is much smaller than that needed for the spiral springs in the portable galvanometer, so that the current needed to turn the coil is correspondingly smaller.

Furthermore the pointer can be dispensed with, and a small mirror mounted on the coil instead. A beam of light from a lamp is focused by a lens on the mirror and then on a scale which may be a metre away. As the coil and mirror rotate, the spot of light on the scale moves. This gives the effect of a long pointer without any weight. Moreover the sensitivity is doubled by the fact that if a mirror rotates through a given angle a beam of light reflected from it rotates through double the angle. Currents of only a few thousandths of a microampere can be detected and, by using suitable scales, be equated in terms of light reflectance from flours or heat produced in bacterial growth, etc.

AMMETERS

An ammeter is much less sensitive than a galvanometer to the flow of electrical current. In the moving iron type, a strip of soft iron is caused to move in the magnetic field produced by a current flowing in the coil around it. The accuracy is limited because of the friction involved and a more accurate ammeter is the incorporation of a permanent magnet, between the poles of which a coil is pivoted which carries the current to be measured. In each type of instrument a pointer attached to the moving portion moves over a graduated scale. One ampere is approximately equivalent to the flow of 6×10^{18} electrons/second.

Ammeters should be placed in *series* with the equipment through which the current is to be measured. They will therefore need a very low resistance compared with the rest of the circuit so that they do not unduly alter the resistance of the whole and produce a false reading.

Current density is a measure of the efficiency of electrical equipment. An ammeter will also indicate the dangers of overloading a circuit. Bulk production can be made efficient if overloaders are installed. An overloader device will cut off electrical power without damage to machinery if the power surges to a high level. Fuses would not be needed on individual machines and within seconds production can be under way. Inspection of overload relays is needed after a day's run to ensure that they will function correctly on the following day's production.

VOLTMETERS

A voltmeter, as its name indicates, is an instrument for measuring the difference of electrical pressure between the two points of a circuit to which it is connected. It should be placed in *parallel* with the apparatus whose potential difference or voltage is required. A voltmeter has to have a high resistance compared with the resistance offered by the circuit so that it will take negligible current, but register a correct voltage for the circuit.

EXERCISE 12

1. How can a galvanometer detect electric current? Describe its function in the flour colour grader.
2. Describe an experiment to produce electricity using
   (*a*) a magnet; or
   (*b*) two different metals.
   Indicate any instruments that are used in a food laboratory which are based upon these experimental observations.
3. Show how a knowledge of Faraday's Law of Electromagnetic Induction has been commercially adapted in the production of a food product.
4. Write brief notes on
   (*a*) resistors; and
   (*b*) rheostat.
   Indicate one use of both in either the butchery or bakery side of food production.
5. Explain the difference between wiring a circuit in series and in parallel. What dangers could result if the incorrect system was used in bulk production?
6. Why is the joule more important than the calorie in the modern food complex? Illustrate your answer with any observations you have made on factory visits.

7. What is the essential difference between an ammeter and a voltmeter? How could they be used to calculate
   (*a*) heat; and
   (*b*) power needed to process a particular food item?
8. Give two advantages of a moving spot galvanometer over a simple galvanometer. How is a galvanometer of use in the quality control of manufactured products?

13 Staining reagents and procedures

FIXED STAIN SMEARS

The range of micro-organisms in the food industry is extensive. Some manufacture chemicals that can counteract the harmful effects of dangerous microbes, others convert the juices of grapes into wine and sweet apple cider into vinegar. The characteristic flavours and textures of pickles, kraut, yoghourt and other foods are related to their metabolism. Although the list of important products resulting from their activity is long, it was not appreciated, even fifty years ago, how dependent man is upon them.

The organic material in dead plants and animals, which is present as carbohydrates and proteins, is broken down by micro-organisms into the simple compounds of which they are constituted. Food is thereby provided for new plants, and subsequently for animals. Bacteria which are present in the nodules of leguminous plants (beans and peas) are able to 'fix' nitrogen, that is to remove it from the air and compound it with other elements to produce compounds from which higher plants can obtain the nitrogen necessary for their protein framework.

Bacteria are of particular interest to the food microbiologist. Their identification often depends upon the *correct* identification of some of their morphological features. To this end the use of certain stains or dyes which are taken up by particular sections of an organism is encouraged. Two advantages are obtained from a staining procedure: the cells are more easily visible after they are coloured, and any differences between cells of different species or within the same species can be demonstrated by the use of relevant staining solutions. This technique is called differential staining.

The essential steps in the preparation of a fixed, stained smear are (*a*) the preparation of a suitable film or smear, (*b*) fixation, and (*c*) the application of one or more prescribed staining solutions. The following procedure indicates how flagella upon a bacterial cell can be identified.

Agar jelly is used as a culture medium and the bacteria allowed to grow for a period of 24 hours at about 37°C. A portion of the growth is then transferred into a sterile glass container with a few millilitres of distilled

159

water. One small drop of the suspension is then placed onto a perfectly clean grease-free slide. The smear is allowed to dry and then 'fixed' by passing the slide, smear down, twice through the bunsen flame.

A number of staining reagents are then prepared, including potash alum and basic fuchsin. These reagents must be added in a definite order at set periods of time. Gram's staining method which is described in detail (see Gram stain) illustrates these points. A time of approximately 10 minutes is recommended for flooding the slide with staining solution for flagella observation. After staining, the slide should be washed with tap water and dried. It is then examined direct or mounted by adding a small drop of mounting fluid (Canada balsam in xylol) on the glass slide and placing it on a cover glass.

Aniline dyes which dissolve in water or alcohol are frequently used as staining solutions. Examples include methylene blue, crystal violet, safranin and basic fuchsin.

SIMPLE STAINING

When bacteria or other micro-organisms are stained by a single solution to a fixed smear the process is one of simple staining. In general the fixed film is flooded with the stain for a fixed period of time, after which the solution is washed off with water and the slide blotted dry. Usually the cells will stain uniformly. Certain bacterial cells when treated with methylene blue show some granules in the interior of the cell to be more deeply stained than the rest of the cell. This shows that specific portions of the two organisms have more affinity for the dye than has the cell as a whole.

Methylene blue can be used to determine the viability of yeast. This is an important factor since it describes yeast activity which determines gassing power. Viability depends to a large extent on the proportion of active (living cells) in the yeast. It is possible, using a very dilute solution of methylene blue, to stain the dead cells in a suspension of yeast.

For ordinary work a quick examination and a rough estimate of the proportion is adequate but exact details can be obtained using a haemocytometer. This is a special slide and cover slip developed for the examination of blood cells. It provides a number of squares into which the cells fall, and which allow sample counting. The test is carried out by placing a few grammes of yeast in a test tube with a little water and shaking well. A few drops of the suspension are then placed on a white tile and mixed well with a few drops of 0.5 per cent methylene blue. A drop of the stained mixture is placed on a glass slide, covered with a cover slip and examined under the microscope. All the dead cells will stain blue and a rough estimate of the percentage of dead cells present can be made. The experiment can be repeated with other yeast samples. A comparison between fresh yeast and old and dried yeast reveals many differences. Figure 33 shows the results on three samples and also a drawing of the structures revealed by staining of a flagellated bacterial cell.

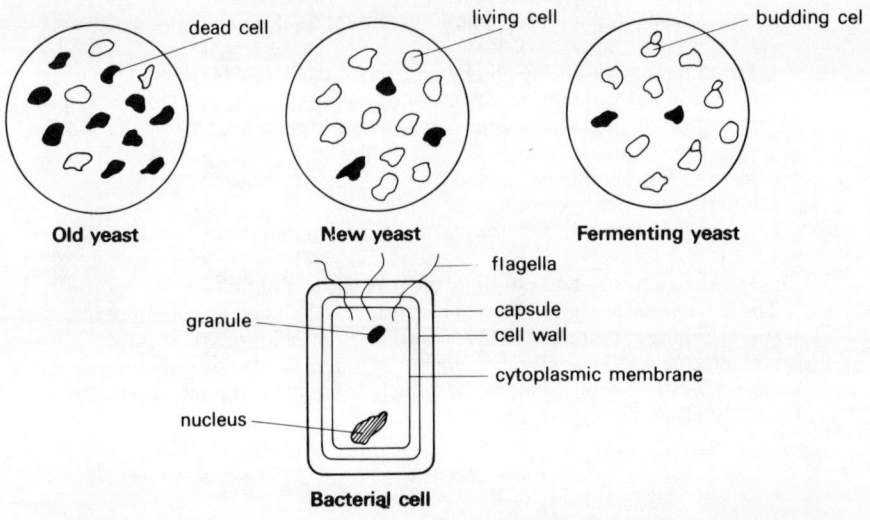

Fig. 33.   Staining of structures.

DIFFERENTIAL STAINING

As the name implies differential staining is concerned with using staining procedures designed to show the specific differences between bacterial cells, or between the different parts of an individual bacterial cell. The techniques involved are more elaborate than those for simple staining in that the cells may be subjected to more than one dye solution or staining reagent.

Usually crystal-violet stain is applied to the fixed smear for about 1 minute, followed by a wash in tap water. Iodine solution is then applied for 1 minute, after which the stain is well washed with water and dried. Alcohol is then added for 30 seconds as a decolouriser and the slide blotted dry. Finally, safranin stain is applied for 10 seconds, followed by a final wash in tap water and a final dry.

This procedure will differentiate between two large bacterial groups, Gram-positive and Gram-negative bacteria. The Gram stain technique is fully discussed in the next paragraph. Ribonucleic acid (RNA), an essential unit in all life processes, can be found in the cell structures of bacteria. Gram-positive bacteria have magnesium in the RNA, but in Gram-negative bacteria magnesium is absent. In both cases the RNA occurs in the cytoplasm.

GRAM STAIN

This is one of the most important and widely used staining techniques in microbiology. The technique takes its name from Hans Gram (1853–1938), a pioneer in the field of bacterial research.

In this process the fixed bacterial smear is treated with various solutions in the following order:

1. Flood the smear on the slide with a basic dye such as gentian violet or methylene blue, and allow to act for 30 seconds.

2. Wash off the stain with Gram's iodine solution and allow the iodine to act for 1 minute.

3. Wash off the iodine with alcohol and leave for 10 seconds.

4. Wash the smear carefully with water.

5. Flood the slide with the counterstain safranin solution, and allow to act for $1\frac{1}{2}$ 2 minutes.

6. Wash with water and carefully blot dry with a filter paper.

The slide is taken and a drop of immersion oil placed on the smear. It is then examined under ×1000 magnification. Figure 34 shows the effect produced. Gram-positive bacteria retain the original basic dye whereas Gram-negative bacteria, which lose the original stain, are counterstained by the safranin.

Operations	Gram-positive bacteria	Gram-negative bacteria
1. Gentian violet	cells absorb colour	cells absorb colour
2. Iodine solution	cell surface contains RNA, protein and magnesium. This complex with iodine reacts with the gential violet and colour is *fixed* in the cell	cell surface lacks magnesium and the gentian violet is *not fixed* in the cell
3. Alcohol	the gentian violet is not dissolved by the alcohol. Cells retain the violet colour	the alcohol dissolves the gentian violet and the cells are decolorised
4. Safranin	as the cells are already stained by the gentian violet no effect is possible with safranin. Therefore the cells remain violet in colour	safranin will stain the decolorised cells which take on a red colour

Fig. 34.   Gram's staining method.

It is not fully clear why some bacteria retain the blue or violet dye whilst others reject the initial dye for the red safranin. However, as the diagram shows the internal difference in cell make-up is obviously involved.

The importance of the Gram stain lies in an examination of the smears stained by this technique. Morphological features and Gram reaction are revealed which are basic to the final identification of the strain involved and its effect upon food and organisms which consume it.

There are three well-established types of bacterial food poisoning. Salmonellosis is the most common, though fortunately it rarely results in death. All three types will indicate their presence by inflicting a mixture of fever, vomiting, headache and diarrhoea. When subjected to a Gram stain test all species of bacteria which cause salmonellosis give a Gram-negative result. This immediately distinguishes it from staphylococcal infection and botulism which both show a Gram-positive reaction.

Morphologically, staphylococcal bacteria are found in irregular clusters whilst the much more dangerous clostridium bacteria which produce botulism are rod shaped. An examination of fresh cultures from suspected cases is valuable in providing health authorities with information vital to the control of infection.

It is important to observe the effects of Gram staining on new cultures for the following reasons. Gram-negative organisms are always constant in their colour retention by the Gram staining method. However Gram-positive organisms can, under certain conditions, show a variation in response. As an example, old cultures of certain Gram-positive bacteria lose their affinity for crystal violet and absorb and retain the counterstain safranin. The environmental background may also produce a change in the organism, as may errors in the staining procedure (for example, adding the various reagents out of order).

Gram-positive bacteria also differ from Gram-negative bacteria in respects other than staining reaction. Penicillin is a powerful antibiotic but it has more success against Gram-positive than Gram-negative bacteria. The opposite effect is noticed when both groups are subjected to mechanical treatment and exposure to some enzymes. Further research has indicated additional differences, the pattern of which shows a fundamental difference between these two groups of bacteria.

ACID-FAST STAIN

The acid-fast stain is another type of differential staining procedure. A reaction is observed when the following solutions in the order listed are applied to a fixed bacterial smear: carbol fuchsin (heated), acid alcohol and methylene blue. The practical procedure is as follows.

1. Flood the fixed smear with carbol fuchsin and *gently* heat the slide (slow steaming of the solution should occur) for a period of 3–5 minutes.
2. Wash well with tap water.
3. Add acid–alcohol (95 per cent alcohol/5 per cent dilute hydrochloric acid) for decolorisation purposes, until only a trace of red dye solution remains.
4. Wash well with tap water.
5. Counterstain with methylene blue for 30 seconds.
Bacteria that retain the carbol fuchsin and appear red are classified as acid-fast; they are termed non-acid fast if they are decolorised by the acid-alcohol and counterstained by the methylene blue.

The presence of a large amount of fatty and waxy substances in the cells of acid-fast organisms cause them to retain the carbol fuchsin. Once the use of heat causes the dye to penetrate the cell it is not readily removed by the acid-alcohol-decolorising agent. Bacteria which do not possess this specific chemical composition will be decolorised and counterstained blue. It has been observed that although acid-fast bacteria are Gram-positive, *not* all Gram-positive bacteria are acid-fast.

Tuberculosis is caused by the tubercle bacillus (*Mycobacterium tuberculosis*). Infection may occur from the ingestion of milk or products from infected cows. The handling of diseased animals and airborne contamination can also produce symptoms of wasting as the central nervous system, bones, joints, kidneys and lymphatic glands are affected.

Bacteria of the genus *Mycobacterium* are acid-fast. The microscopic examination of specimens suspected of containing the tubercle bacillus using the differential staining technique of acid-fast stain will provide a clue as to the presence or otherwise of possible infection. Only those organisms that stain red will merit further work on tubercular lines.

Fortunately tuberculosis is being controlled in animals destined for human consumption. Increased standards of hygiene and pasteurisation of milk and its products have contributed greatly to the dramatic fall in cases since the turn of the century. Inspection of meat carcasses, particularly pigs, is standard procedure. It has been stated that it is often difficult to find actual specimens to show tuberculosis to trainee meat inspectors, which is gratifying to the public at large.

Brucellosis or undulant fever is more common, but this bacterial infection does not present the industry with the dangers inherent with tuberculosis. Differential staining followed by specific tests for both bacilli will present a rapid identification.

CELL STRUCTURE STAINING

The individual structures which are interior or exterior to a bacterial cell will require special staining procedures in order to identify them. Because of the similarity of individual species this work is very important if pathogenic bacteria are not to be confused with their harmless counterparts.

The differential staining procedures used all adhere to the principles laid down for the Gram and acid-fast stain techniques; namely, the structure in question should exhibit a different affinity towards a particular dye from that of the rest of the cell, making differentiation possible.

A further aid to cell structure details is the negative staining technique. It is possible that the fixing of bacterial smears prior to staining and their subsequent exposure to a series of reagents may result in some distortion of the cells. Microscope examination could therefore lead to incorrect and even harmful conclusions. This is not likely to occur in the negative staining procedure. A bacterial suspension is prepared for negative field staining by mixing it with Indian ink or nigrosin and then spreading it as a thin film across a slide and allowing it to dry. The bacteria will appear transparent when outlined by the dark background. In some respects the preparation is comparable in appearance to a dark-field preparation, since the organisms appear light in a dark microscopic field. Negative staining is advantageous for the study of bacterial morphology because the bacterial cells are not subjected to a vigorous physical or chemical treatment.

BACTERIAL COUNTING METHODS

It is often essential to prepare a report on the size of the bacterial population in a food sample. The problems associated with food spoilage and food poisoning are always factors for concern. Four techniques are available: (*a*) viable count; (*b*) roll tube; (*c*) surface or drop count; and (*d*) membrane filter count.

The viable count is the most accurate method. It is assumed that a visible colony will develop from each bacterium. However, bacteria are often clumped together and therefore a single colony may develop from more than one original organism. Since any agitation, as in the preparation of dilutions, will break up or cause the formation of clumps it is difficult to obtain a result which is entirely free from experimental error. The figures obtained from a single test may be erroneous, and therefore a large number of tests are required.

Initially pipette 9 cm^3 of 25 per cent strength Ringer's solution into sterile cap bottles; these are dilution blanks. When diluting liquids, e.g. milk, for counts proceed as follows. Mix the samples well by shaking. Use a sterile straight sided pipette dipped about 1 cm below the surface and remove 1 cm^3. Deliver into the first dilution blank about 1 cm^3 above the liquid level. Wait three seconds to drip out and then discard the pipette. Continue for the required number of dilutions and remember to discard the pipette after delivering the contents otherwise the liquid on the outside will contribute to a cumulative error. The dilutions will be:

Tube No.	1	2	3	4	5
Dilution	$\dfrac{1}{10}$	$\dfrac{1}{100}$	$\dfrac{1}{1000}$	$\dfrac{1}{10000}$	$\dfrac{1}{100000}$
Volume of original fluid/cm^3	0.1	0.01	0.001	0.0001	0.00001 cm^3
or	(10^{-1})	(10^{-2})	(10^{-3})	(10^{-4})	(10^{-5}) cm^3

Proceed now to a plate count. Melt the nutrient agar and place 10 cm^3 in each tube. Cool in water bath to 45°C. Set out petri dishes for dilutions under test and label with dilution number. Pipette 1 cm^3 of each dilution into the centre of the dishes. Use a fresh pipette for each dilution. Do not leave the dish uncovered for longer than is absolutely necessary. Add the contents of one agar tube to each dish in turn and mix by moving the dish six times in a clockwise circle of diameter about 15 cm. Allow the medium to set, invert and incubate for 24–48 hours at 37°C. To count select plates showing between 30 and 300 colonies. Use a colony counting box. Place the dish bottom side up and divide into sections by ruling through the diameter with a fine felt pen or wax pencil. Count by marking the bottom of the plate with a wax pencil above each colony.

The final stage is to calculate the precise number of bacteria present per gramme in the original sample. For example if a 10^{-4} dilution had 114 colonies present it can be assumed that 1 cm^3 of that dilution contained 114 cells. As the 10^{-4} dilution has only one part in 10 000 of the original sample it follows that the raw material must have 1 140 000 cells/gramme.

The following formula can thus be used.

$$\text{Bacterial population (cells/cm}^3 \text{ or cells/g)} = \text{Colony count} \times \text{dilution factor}$$

In all cases where the dilution is given as 10^{-n} add n zeros after the number of colonies observed for original population of bacteria.

The roll tube count involves the use of bottles containing media rather than plates. Inoculate the bottles with the diluted material and rotate horizontally until the medium sets. After incubation the colonies are counted. The tube medium (2–4 cm^3) is placed in screw capped bottles, 0.5–1.0 per cent more agar than usual is required. Melt the agar and cool to 45°C in a water bath. Add 1.0 cm^3 of each dilution and rotate horizontally in cold water until the agar is set in a uniform film around the walls of the bottle. This requires some practice. An alternative method is to roll a tube on a block of ice. Invert the roll tube cultures and incubate so that the condensation collects in the neck. To count draw a line parallel to the long axis of the bottle and rotate the bottle, counting the colonies on a roll tube counting box with side illumination.

In the surface or drop count method small drops of material are placed on agar plates and after incubation the colonies are counted in the inoculated areas. Pipettes known as '50 drop pipettes' are used. The delivery of this pipette is equal to 50 drops/cm^3. Alternatively hypodermic needles can be used (gauge 19). At least 5 drops from each dilution are added from a height of not more than 2 cm to avoid splashing on the culture plates. Replace lid but do not invert until the drops have dried. After incubation select those plates showing colonies in the drop areas. Count the colonies in each drop. Divide the total count by the number of drops counted and multiply by 50 to convert to 1 cm^3 and by dilution used.

The membrane filter count involves passing the liquid containing the bacteria through a filter which will retain the bacteria. Allow the filter to absorb the culture medium, incubation occurs and colonies are produced which can then be counted. The filters are usually made of a thin porous cellulose ester composition with a pore size of 0.5–1.0 μm diameter enlarging to 3–5 μm at the bottom. Bacteria are held back at the top but the culture medium can easily rise to them by capillary action. A grid to facilitate counting is ruled on the upper surface of each filter. The filter apparatus is made of metal and consists of a lower funnel which carries a carbon-sinter platform surrounded by a silicone ring and the flange of the upper funnel. The upper and lower funnels are held together by a clamping ring.

The first step is to pipette 25 cm³ of the medium onto an absorbent pad in an incubation container. This should wet the pad only as far as the edges. The filter carrier is erected over a filter flask connected to a pump by a non-return valve giving a suction of 25–50 mm mercury. Check the filter is in place and tighten the clamping ring. Pour the fluid to be examined into the upper funnel and apply suction. When filtration is complete restore pressure and remove filter with sterile forceps. Apply it to the surface of the wet pad in the incubation container. Put the lid on the container and incubate. For total aerobic counts use tryptone soya membrane medium.

To count coliform bacteria place the filter on Resuscitation medium for 1–2 hours then on Mac Conkey membrane medium, otherwise counts will be too low; count in oblique light. The filter can be stained in 0.1 per cent methylene blue for 30 seconds and then placed on a pad saturated with water. Report as membrane colony count per standard volume.

It is important to note that despite the error that may arise through cell clumping the plate count is one of the best ways of estimating the characteristics of a bacterial colony. Not only is it quantitative but by microscopical examination of the colony types present a good idea of the individual strains present may be obtained.

EXERCISE 13

1. Describe two procedures whereby specimens can be prepared for microscopic examination. What are the advantages of each?
2. What factors distinguish a differential staining procedure from a negative staining technique?
3. Explain the significance of Gram's staining method to the milk industry.
4. What generalisation can be made about Gram-positive organisms and their reaction to the acid-fast strain? Give any applications that may be of interest to a cattle breeder.
5. Define the term cell structure staining. Show how it can contribute to the production of good bread under different environmental conditions.
6. Which staining procedures would be important in the control of infection in poultry houses? Suggest briefly how the information obtained would be put to a practical use.

14 Microbiological media

NUTRITIONAL REQUIREMENTS

The satisfactory study of micro-organisms depends upon an ability to cultivate them under laboratory conditions. A knowledge of their food material and of the physical conditions needed for optimum growth is therefore essential. Research has resulted in the production of various media for their cultivation.

The composition of these media is variable due to the different nutritional requirements of bacteria. Bacteria also show great differences in ideal growth temperatures. For example a few will grow well at 10°C, a much larger group will prefer 45°C and a tiny fraction may even grow well at 70°C. Some bacteria will require atmospheric oxygen (aerobic conditions), others are inhibited by it (anaerobic conditions). Their metabolism will also be affected by pH, indicating a preference for either acid or alkaline conditions.

All living organisms need a source of energy for their growth. Green plants are capable of utilising radiant energy and are known as phototrophs. Other organisms which cannot use radiant energy direct (for example, animal life and non-green plants) rely upon the reactions of chemical compounds to provide energy. These are known as chemotrophs. Both nutritional types are found in the bacterial world.

Carbon is required by living organisms. It is supplied either as carbon dioxide or in a more complex form, such as sugars or other carbohydrates. The mechanism of photosynthesis is used by plants to convert carbon dioxide into carbohydrate. Animals depend upon plants to provide the carbon they need in a suitable form of carbohydrate. Certain bacteria can use carbon dioxide direct, others require more complex organic compounds of carbon. Those types of bacteria which rely solely upon carbon dioxide as their only source of carbon are referred to as autotrophs whereas bacteria that require an organic form of carbon are termed heterotrophs.

Nitrogen is needed by living organisms in some form, ranging from the free element to complex protein molecules. Plants absorb nitrogen in the form of soluble inorganic salts, such as potassium nitrate, by the process of osmosis. Animals require nitrogen as the degradation products of protein molecules or as proteins themselves.

In the process of digestion the amino acids are assimilated by the body chemistry into forms suitable to build new tissue. Bacteria are very versatile in their nitrogen outlook. Some types use atmospheric nitrogen, others need inorganic nitrogen compounds. A number are also able to derive nitrogen from proteins or practically any naturally occurring organic compound containing nitrogen.

Bacteria, in common with other living organisms, need water to grow; and, like plants, they require nutrients in solution before they can enter the organism. Metallic elements also are needed by living organisms for normal growth, and these include sodium, potassium, calcium, manganese, magnesium, zinc, phosphorus, copper and cobalt. Bacteria are no exception, but need only mere traces to aid growth.

Bacteria, unlike the higher animals including man, are able, in a large number of cases, to synthesise their vitamin requirements from simple compounds present in a nutrient medium. Others cannot grow unless one or more of the vitamins are contained in the medium.

Heterotrophs are of most direct concern to man, particularly in their relationship with food. This group contains all the bacteria that can produce diseases in animals and plants. More attention has therefore been directed towards the development of media for the cultivation of heterotrophs rather than autotrophs.

AUTOTROPHIC BACTERIA

The preparation of suitable growth media for this group is a relatively simple process; only inorganic compounds are needed. As an example a medium comprising powdered sulphur, calcium chloride and potassium dihydrogen phosphate as the main chemicals with traces of ammonium sulphate, magnesium sulphate and ferrous sulphate dissolved in water is suitable for the growth of sulphur-oxidising bacteria. The total weight of chemicals in 1000 g of water is only about 25 g and yet this will provide all the essentials needed for growth. This illustrates that autotrophs have an elaborate capacity for synthesis and are able to transform these compounds by synthesis into new living cells measuring only a few micrometres in length.

Other autotrophic bacteria need a similar inorganic medium nutrient to the one described but differ in that certain other specific chemical compounds have to be substituted or added. A medium which is made from known chemical compounds is referred to as a chemically defined or synthetic medium.

HETEROTROPHIC BACTERIA

Depending upon the species, heterotrophic bacteria may have simple or relatively complex nutritional requirements. Heterotrophs, unlike autotrophs, cannot use carbon dioxide as their sole source of carbon but need one or more organic compounds. This can be illustrated by reference to the nutrient media required for *Escherichia coli* (autotroph) and *lactobacilli* (heterotroph). The former medium, apart from glucose, contains only inorganic salts whilst the latter includes at least ten complicated vitamin substances represented by riboflavin, niacin and folic acid, etc.

The routine cultivation in the laboratory of heterotrophs is not based upon a complex mixture of pure chemical compounds but rather upon using complex raw materials, such as peptones, meat and yeast extracts. This combination will produce media capable of supporting a wide variety of organisms. Agar is included as a solidifying agent when a solid medium is desired. Nutrient broth and nutrient agar represent a relatively simple liquid and solid medium which can support the growth of many common heterotrophs. Approximate compositions of these media are as follows:

Nutrient broth		*Nutrient agar*	
Beef extract	2.5 g	Beef extract	2.5 g
Peptone	5.2 g	Peptone	5.2 g
		Agar	14.5 g

In both cases the mixtures are added to 0.001 m³ of water. About 0.005 g/m³ of yeast extract may be added to the above formulae to improve the nutritional quality of the medium, as yeast extract contains several of the B vitamins.

TYPES OF MEDIA

A selected number of heterotrophs will not grow well on nutrient agar or nutrient broth and a few will actually fail to grow. *Proteus vulgaris* is found in water, sewage and decayed matter and will obtain its energy from the oxidation of sugar and other organic compounds. Its nutritional requirements include the B vitamin niacin which is not present in agar. The agar is a gelatin-like material obtained from certain seaweeds and related to the carbohydrates.

Organisms such as *Proteus vulgaris* are designated fastidious heterotrophs. Other organisms may need specially prepared media to enable them to be isolated and identified. Figure 35 shows some of the complex materials used as ingredients of media for the cultivation of heterotrophs.

Media may be classified on the basis of their physical state. Solid media such as slices of potato could be used for special cultivation of bacteria. Nutrient agar would represent a solid which can easily be reversed to a liquid; it will gel and set at room temperature. Semi-solid media which contain only about 0.5 per cent or less of agar would set to a 'custard-like' consistency. Nutrient broth would typify a liquid medium.

A medium which contains plant extracts, animal tissues or blood

Raw material	Composition of raw material	Importance to bacteria
Beef extract	lean beef tissue as an aqueous extract and concentrated to a paste.	provides carbohydrates, organic nitrogen compounds, salts and water-soluble vitamins.
Peptone	obtained by acidic or enzymatic digestion of proteinaceous materials such as casein, gelatin and meat.	provides organic nitrogen and some vitamins. Carbohydrates may be produced. It depends upon the nature of protein source.
Agar	complex carbohydrate from marine algae which is processed to remove any extraneous matter.	no nutrient supplied but used as a solidification agent for media below 45°C.
Yeast extract	supplied as a powder and prepared from aqueous yeast cells.	rich source of B vitamins.

Fig. 35.   Materials in bacterial media.

added to a nutrient broth or agar will be enriched. These additional nutrients would be essential for the growth of certain fastidious heterotrophs.

Certain chemical substances added to nutrient agar will prevent the growth of one group of bacteria without inhibiting others; such a medium would be selected. An example would be the use of crystal violet which, at certain concentrations, prevents the growth of Gram-positive bacteria without altering the growth of Gram-negative bacteria.

When blood is added to an agar medium, and a mixture of bacteria is inoculated on to it, some of the bacteria may destroy the red blood cells whilst others do not. The destructive bacteria are hemolytic in nature, while those which have no observed effect are non-hemolytic.

Various media of prescribed composition are used to determine the bacterial content of milk and water; tests for the *Staphylococcus aureus* and coliform bacteria population in milk are routine.

When testing disinfectants against pathogenic bacteria assay media are prepared. The destructive effects can then be visually seen using plates of agar jelly. Other media can give numerical data on amino acids, antibiotics or vitamins in food at any stage of production.

Finally a wide variety of media are available to determine the type of growth produced by organisms and also to investigate their ability to produce chemical changes.

PHYSICAL GROWTH CONDITIONS

After the preparation of a suitable nutrient medium for a particular bacterial strain the reaction of the organism to its environment must be studied if optimum growth is to be attained. A most important variant is temperature. The rate of growth, total growth, morphology and metabolism of the organism are all related directly to the prevailing temperature. Every species of bacteria will grow best over a certain temperature range, and bacteria can be divided into three distinct groups on the basis of these ranges.

(a) Psychrophiles—those that grow best at temperatures below 20°C.

(b) Mesophiles—those that grow best at temperatures between 20 and 40°C.

(c) Thermophiles—those that grow best at temperatures between 40 and 60°C.

Certain thermophilic bacteria have a growth range which can extend into the mesophilic region. These are termed facultative thermophiles. Others in the thermophilic category will grow best above 60°C and are unable to grow in the mesophilic range, these being true thermophiles or obligate thermophiles. Unlike an obligate organism, a facultative one has more than one area of growth. In a period of 12–24 hours the temperature of incubation which allows for the most growth is called the optimum-growth temperature.

The gaseous requirements of bacteria with regard to oxygen and carbon dioxide vary from species to species. In particular, bacteria show a wide response pattern to oxygen. Some need free oxygen, others will only grow in its complete absence, whilst there are intermediate types which can grow with or without oxygen. Four distinct groupings are possible:

(a) aerobic—grow only in the presence of oxygen;

(b) anaerobic—grow only in the absence of free oxygen;

(c) facultatively anaerobic—grow in the presence or absence of free oxygen; and

(d) Microaerophilic—grow in the presence of traces of free oxygen.

Figure 36 shows how each type would grow when inoculated into an agar medium. The cotton-wool plug serves as a restrictor to other airborne bacterial organisms.

The cultivation of anaerobic bacteria needs special attention and equipment. The medium must not contain any atmospheric oxygen. This can be achieved chemically by the addition of a suitable reducing compound such as sodium thioglycollate to the medium. A physical alternative is the mechanical removal of oxygen from an enclosed vessel which contains the inoculated medium. The air is removed by a pump and replaced by nitrogen.

A third major physical factor is pH. The optimum for most bacteria lies between 6.5 and 7.5 which produces either mild acidic, neutral or mild alkaline conditions. Most species will have a growth range of pH 4–9, with a few bacteria showing maximum growth at the extremes of this range. It is also observed that when bacteria are cultivated in a medium adjusted to a given pH it is likely that the pH will change, due to substances produced by the organism which may be either acidic or basic.

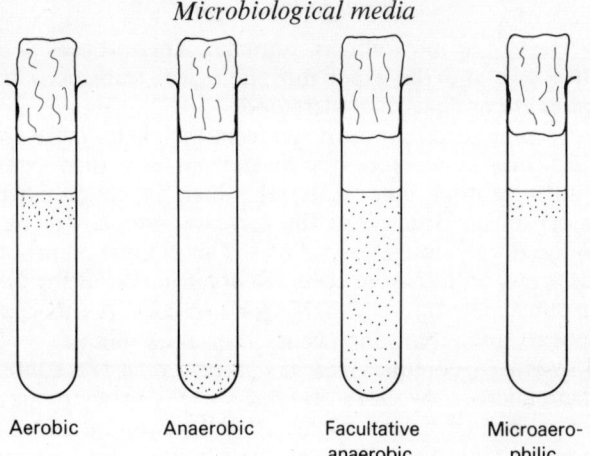

| Aerobic | Anaerobic | Facultative anaerobic | Microaero-philic |

Fig. 36.   Oxygen effect upon bacterial growth.

The change in pH could be great enough to inhibit the growth of the organism. However, the introduction of a buffer into the system will prevent any radical shift in the pH. Compounds of the type KH_2PO_4 and K_2HPO_4 either used individually or collectively are widely used in bacteriological media for this purpose. Peptones also possess some buffering capacity.

Additional physical factors such as light and salt will affect bacterial growth. Photosynthetic autotrophic organisms need light as a source of energy. Bacteria which are found in brines or salt packs can be a source of food spoilage. These halophilic (salt liking) bacteria are obligate in character because they only grow in a medium of about 10–15 per cent salt concentration. A relationship exists between halophilic bacteria and the peculiar osmotic pressure effect of these organisms. Usually a high osmotic pressure externally would dehydrate a bacterium. This illustrates again the versatility of Nature in producing strains capable of surviving under extreme conditions.

REPRODUCTION

When bacteria alight on a suitable food supply or are inoculated into a suitable medium under correct conditions a tremendous increase in numbers occurs in a very short space of time. Some species achieve a maximum population within 24 hours, others may need a longer period of incubation for peak performance. Growth, when applied to micro-organisms, including bacteria, refers to any changes that occur in the whole population rather than to an individual organism.

The most common process for the increase in a bacterial population is binary fission, sometimes known as transverse fission. In this process the single cell divides into two, after the development of a transverse cell wall to separate the intracellular contents. This process is an asexual reproductive one and the exact morphological transitions involved in binary fission are not clearly understood.

It has been suggested that nutrients from the medium are absorbed by the cell. Enzyme systems of the bacterium can then convert these nutrients into protoplasmic material which is characteristic of the particular organism. Because of the increased amount of nuclear substance produced, cell elongation follows. This elongation process is often more evident in bacilli than in cocci. Reorganisation of the cell contents now occurs and material is available for two cells. A transverse wall or septum appears and separation occurs after a few minutes.

Special staining techniques and new microscopic procedures, particularly electron microscopy, have revealed some of the cytological changes that occur in binary fission. Some of the major questions posed by bacterial reproduction are as yet unanswered. For example, what initially stimulates cell division? How is the septum produced? Which forces control the organisation of the newly formed protoplasmic material within the cell?

These questions can also be asked for cellular division in higher plants and animals. One can assume that the underlying principles are similar in all biological systems. An understanding of these processes as they apply to bacteria can have important applications for all cells. Cytology is important in producing animals with the best potential for breeding purposes. Selected animals can produce offspring with a high resistance to certain infections. Milk yield can rise and weight increase in cattle, both valuable economic considerations to the farmer; all of which can be traced to the original cell structure.

A sexual mode of reproduction is also possible with certain species. With *E. coli* two different strains, each having identifiable characteristics and differing in minor details, were grown together. After incubation the cells were plated and then tested for the characteristics of the 'parents'. The result was that a small but significant number of the new isolated strains possessed characteristics of both parent types. One feature of the new strain was its ability to ferment lactose and also to be resistant to the antibiotic streptomycin. Neither parent cell possessed this ability. One parent cell would not ferment lactose but was resistant to streptomycin, the other fermented lactose but was extremely sensitive to the antibiotic.

Further investigation revealed a second new strain which would neither ferment lactose or withstand streptomycin. It must be stated that the majority of the new cells investigated resembled one or other of the parent types. However, the minor occurrence of characteristics from both parents in the daughter cells can best be explained by the fusion of cells (sexual reproduction) as the process is analogous to higher animal reproduction.

Actinomycetaceae are an order of bacteria which are concerned with the degradation of complex substrates such as proteins. They are found in soil and have the ability to reproduce by the formation of a filamentous growth followed by fragmentation into small units which can then develop into cells of normal size. During the latter stages of growth the production of a mycelium prior to fragmentation is similar to the formation of a mould.

Again, other bacteria which are parasitic on fresh-water crustacea (*Hypomicrobiales*) are able to reproduce by a budding process. An outgrowth, or bud, develops from the parent cell and, after enlargement, separates from the parent as a new cell. Budding is characteristic of yeasts, rather than bacteria.

GROWTH RATE

Binary fission has already been indicated as the prevailing process of bacterial reproduction; one cell divides into two new cells. Thus from a single cell, the increase in population is by geometrical progression $1 \rightarrow 2 \rightarrow 4 \rightarrow 8 \rightarrow 16 \rightarrow 32 \rightarrow 64$ etc.

The time taken for a newly divided cell to grow and divide again is known as the generation time. This factor is vital in the rate of growth of a bacterial culture. Bacteria vary in the generation time. For *E. coli* it is about 20 minutes, but for others it may be several hours.

The generation time will be a variable factor for a particular bacterium as environmental conditions change. Factors such as the amount and kind of nutrients available, temperature and pH can all alter an individual generation time. Because the population doubles every time a cell divides the growth of pure cultures of bacteria is plotted graphically as the logarithm of the number of cells against the time of incubation.

There appear to be three distinct phases in the growth of a microbial population. Phase one is the log phase in which the cell increases in size, content of nuclear material and in the content of certain enzyme systems. At this stage the number of cells produced is not very great. This is quickly followed by a second phase of logarithmic growth during which the cells continue to grow and divide at a constant rate. Phase three is now reached; this is a stationary phase, and may be terminated by the bacteria dying slowly or rapidly, depending upon the nature of the organism and the restraining influences.

The growth rate factor for any organism is the basic reference point for any work involving food spoilage and food preservation. One factor which emerges is the success achieved by using antibiotic ice in fish packing. Chlorotetracycline in ice at about 20 p.p.m. will preserve halibut and cod for ten days at 1–3°C. A comparison was made between fish stored in normal ice and antibiotic ice with regard to bacterial growth after 14 days. The bacterial load from fish stored in antibiotic ice was only 12.5 per cent of the figure obtained for normal ice, a dramatic fall indeed.

PURE CULTURE TECHNIQUES

The variety and number of micro-organisms (bacteria, yeasts and moulds) found in food or food containers, etc. present a problem if individual species are to be isolated. A technique often used for collecting micro-organisms is to prepare a number of petri dishes containing nutrient agar which is set up and ready for use. A sterile swab can be taken and rubbed over the working surface, machinery, hair, clothes, etc. after which it is wiped over with the medium in a zig-zag fashion.

The lid is placed on the petri dish which is inverted and then incubated at a suitable temperature (e.g. 30°C) for at least 72 hours. After incubation, the plates are examined for growth. It will be observed that numerous organisms of different types are present, particularly in the bacterial field.

The term mixed culture refers to more than one species of bacteria growing in a medium and is always a characteristic of random swabbing. It will therefore be necessary to obtain a culture which contains only one species—a pure culture.

A number of techniques are available for the preparation of a given pure culture from a mixed population. One of the most popular is the streak-plate technique. Using a transfer loop, a small portion of a bacterial specimen is placed on the surface of solid nutrient agar and streaked as shown in Fig. 37. By drawing a sterilised transfer needle across the surface several times, in the direction A–B, the specimen on the end of the needle is transferred to the agar. The needle is now sterilised in a bunsen burner flame, cooled, dipped into the sample and streaked several times in the direction B–C. This procedure is now

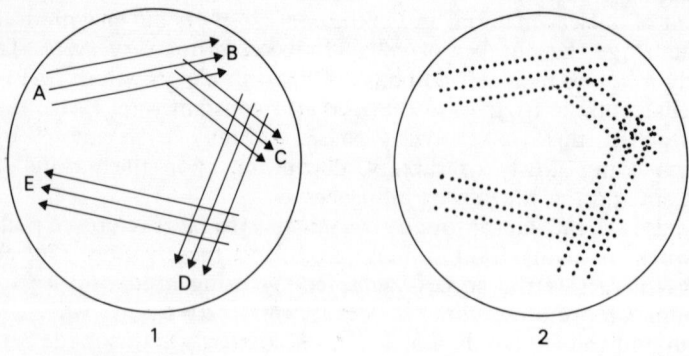

1. streaking procedure
2. plate after incubation period

Fig. 37.   Pure cultures (streak-plate technique).

repeated for directions C–D and D–E. After incubation the colonies will appear as streaks.

A 'thinning-out' of the bacterial population has been achieved and, if properly performed, the bacterial cells will be far enough apart in some areas of the plate to avoid contamination as each cell grows into a colony. Each isolated colony will be the progeny of a single cell and hence a pure culture. With species such as staphylococci and streptococci each group of cells will have its own characteristic shape. By using a fine sterilised needle a portion of a colony can be transferred to a second petri dish containing nutrient agar grown in isolation. The method has now become routine for quality control tests on food but has the limitation that only a small amount of the specimen can be spread over the agar.

When the sample is collected in a test-tube as milk or water would be, it is examined by the pour-plate technique. One loopful of the liquid or suspension is transferred to a tube containing cooled liquid agar. The tube is rolled between the hands to effect adequate mixing of the inoculum and a similar loop transfer is made into two further tubes containing nutrient agar. Each of the three tubes is then poured into separate petri dishes containing nutrient. After incubation the plates are examined for isolated colonies. The one selected has a small portion removed and transferred to a tube of sterile medium for pure culture growth.

A third technique can provide a quantitative as well as a qualitative result with a particular species. Samples of contaminated water may contain coliform type bacteria which can give rise to food poisoning. As the generation time factor is small, the organism being sought might be present in much greater numbers than any other organism. By taking a known volume of the sample and transferring small amounts (about 1 cm^3), using a micropipette, into several culture tubes of nutrient agar, the final tube would be a pure culture of a known dilution. It is again advisable to confirm by a plating procedure the purity of a culture isolated in this fashion.

The growth characteristics of a pure culture may be studied by (*a*) growth on agar slants; (*b*) growth in broth; or (*c*) growth in gelatin stabs.

An agar slant culture is prepared by making a streak inoculation up the middle of a sloping surface of nutrient with a transfer needle. When broth is used, the tubes that contain it may be inoculated with either the transfer needle or the loop. Usually the loop is preferred when the inoculum is a liquid. The inoculum in all cases is the material containing the micro-organisms. For the stab culture, the transfer needle which carries the inoculum is introduced into a straight line from the top to the bottom of the tube and removed along the same path.

An appropriate method is therefore applicable to any sample whose microbiological population merits attention in the interests of public health and food acceptability.

PROBLEMS OF YEASTS AND MOULDS

Yeasts have low heat resistance which virtually eliminates them from spoilage of heat-processed canned foods. Cases which do arise are often traceable to leakage or underprocessing. As a class they are often responsible for changes in food due to fermentation. Products containing acids, sugar and salt as preservative factors are vulnerable to microorganisms. These include jams, acid sauces, sausages, cheese and meat. Their general metabolic requirements include:

(*a*) Abundant moisture, minimum 15 per cent.

(*b*) Mesophilic temperature range, although some are psychophilic, for example wine yeasts.

(*c*) Acid media, pH 4–4.5.

(*d*) Free oxygen (aerobic) although fermentative types can grow anaerobically slowly.

(*e*) Ammonia, urea, amino acids, polypeptides and vitamins.

Oxidative (aerobic) yeasts break down sugars, acids, alcohols and fats to carbon dioxide and water. Fermentative (anaerobic) yeasts break down sugars to ethanol and carbon dioxide.

Although the above requirements are true for the majority of yeasts there are exceptions. Benzoic acid used in permitted levels as a preservative for citrus fruits at pH 3 does not prevent the yeasts *Saccharomyces acidifaciens* or *S. elegans* from growing. The flavour of wine is dependent upon correct yeast selection. Sulphur dioxide in controlled amounts may be used in the initial stages to destroy wild yeasts but will leave the wine yeasts unaffected. Apple juice stored in tanks may produce flavours traceable to minute quantities of acetic or butyric acids. Yeasts are believed to be a causative agent despite the liquid having a pH outside the 4–4.5 range.

Fermentative yeasts are found throughout the medium whereas oxidative yeasts produce films. Film yeasts are characterised by their low gas production and by the formation of esters revealed by their apple-like smell.

True yeasts are sporogenous and produce ascospores. *Pichia* represents an oxidative true yeast and will grow on pickles, cheese and evaporated milk when exposed to the air. *Hansenula* and *Debaryomyces* will oxidise organic acids and spoil strong wines and spirits; the latter will also grow in meat brines. Fermentative true yeasts which can cause trouble include *Hanseniaspora* (fruit juice spoilage), *Mussbaumeri* (spoils jam, honey) and *Pastorianus* (spoils beer).

Wild or false yeasts are asporogenous and no ascospores are produced. Oxidative types include *Candida* (wine spoilage), lactic acid oxidation in yoghourts, fat spoilage in meats. *Trichosporon* will form pink spots on chilled beef and *Rhodotorula* will form red, pink or yellow spots on food, especially meat. Fermentative examples are represented by *Torulopsis sphaerica* (spoils milk and dairy products) and *Kloeckera* (wine spoilage).

Unlike yeasts the problems associated with moulds tend to be more apparent to the general public. Colour changes in food are more noticeable than flavour differences. Like yeasts they need a minimum moisture content of about 15 per cent. Most varieties are mesophilic and grow well between 25 and 30°C. *Mucor, Rhizopus* and *Penicillium* all produce undesirable surface contamination of white, grey or green-blue areas. A small number are psychrophilic and can affect food in a domestic refrigerator if it is well endowed with carbohydrates and at temperatures of 2–5°C. Thermophiles, for example *Byssochlamys*, spoil canned or bottled fruit.

Moulds are aerobic and can tolerate a wide pH range (2–8.5), although the majority are optimum at pH 2–4. *Sporotrichum* is responsible for the 'white spot' of meat; dry soil may contain 120 000 spores/gramme and animal faeces (fresh) 60 000 spores/gramme. A reduction in numbers can be achieved by treating the surface of meat with hot water containing bactericidal/fungicidal detergents.

A few moulds may produce mycotoxins which can be poisonous. Peanut meal used in the feed for calves, pigs and sheep was found to contain a compound, aflatoxin, produced by strains of *Aspergillus flavus* growing on infected peanuts. Damage occurs to the liver cells over a period of time as the toxin is cumulative. The mould grows on the nuts after harvesting, particularly if drying is delayed or the shells damaged. Mycotoxins have also been found in corn and millet; this may be a problem in areas of malnutrition. Rice may be attacked and produce absorbable toxins. The high incidence of liver complaints in rice-eating countries which are low in high protein may be linked with this mould toxin.

Bakeries are potentially places with a high risk of mould infection, but this can be minimised if the correct precautions are taken, for example:

(*a*) Flour bleaching with chlorine or benzoyl peroxide and humidity control during storage.

(*b*) Use of non-toxic bactericides and removal of spore laden dust.

(*c*) Make maximum use of sunlight or artificial ultra violet rays.

(*d*) Loaves must be well cooled before wrapping to reduce condensation beneath the wrapper.

(*e*) Do not store bread in a warm humid place.

(*f*) Add mould inhibitor (sodium propionate) at the rate of 0.1–0.3 per cent of flour weight.

(*g*) Check for contamination of the slicing machine and remember that sliced bread is especially vulnerable.

Control of relative equilibrium humidity and the incorporation of plastic linings or related material are of great value in mould control. Modern packing of fruit produces an enriched carbon dioxide atmosphere and diminishes oxygen availability for respiration purposes. Those features will tend to suppress spoilage and increase profits.

EXERCISE 14

1. Outline a suitable method (stating media and conditions of incubation) for the isolation of a thermophilic anaerobic bacterium in a meat product.
2. How would you examine the bacterial population of
   (a) fresh; and
   (b) old bread?
   Comment on any differences you might find.
3. How are bacteria classified in relation to external physical factors?
4. What bacterial problems are likely to arise in
   (a) refrigeration temperatures; and
   (b) cooking temperatures when applied to meat and meat products?
5. Explain the terms obligate anaerobe and facultative anaerobe. What is the practical significance of these terms to a confectioner?
6. Why may it be necessary for a farm laboratory to maintain a stock-culture selection? How can this be done?
7. Assume that you are examining colonies on a plate which has been inoculated with a specimen of milk. What characteristics might be helpful in drawing a conclusion as to the quality of the sample?
8. A pure culture has been isolated by an acceptable technique. What can be done to prove that the isolate is actually a pure culture?

15 Applied micro- biology and food analysis

FERMENTATION

A large number of chemical reactions which are commonly referred to as fermentations are used in the production of beer, wine, vinegar and antibiotics (for example, penicillin). These chemical changes are brought about in organic substances by living organisms, such as yeast and bacteria, and involve enzyme activity. The classic case of fermentation involves the action of zymase upon certain sugars where ethanol and carbon dioxide are produced according to the equation:

$$C_6H_{12}O_6 = 2C_2H_5OH + 2CO_2.$$

In this process the carbohydrates are subjected to anaerobic oxidation by enzyme action. Gaseous oxygen is not involved in the energy-yielding process. Pasteur's name is permanently linked with the brilliant understanding of the methods and processes necessary to produce good quality wine and beer. He showed that the fermentation of fruits and grain to produce alcohol was the direct result of micro-organisms. By his examination of many batches of 'ferment', different microbes were shown to produce different results. Good and bad quality wine was related to microbiological population. By proper inoculation techniques, a manufacturer can be assured of a good and uniform product.

Pasteur suggested that undesirable microbes might be removed by heat, not enough to alter fruit flavour of the juice, but enough to render the microbes harmless. He found that holding the juices at a temperature of 145°F for about half an hour achieved success.

Modern food industries use the pasteurisation process for a wide range of fermentation activities. A familiar application is its use in the dairy industry where it destroys pathogenic bacteria in the milk.

Fermentation actions are also associated with the production of energy. Glucose can be fermented to produce either acetic acid or lactic acid (depending upon the enzymes present). The energy produced is much less than that produced by direct oxidation. It can be shown that the anaerobic breakdown of glucose (glycolysis) takes place in essentially the same fashion in muscle cells, yeasts or bacteria. A large number of enzymes are involved. The reaction of each enzyme produces a product which is the substrate for the next enzyme in the system. In the living beast, glucose is broken down by glycolytic enzymes into energy which is 'stored' in the phosphate bonds of adenosine triphosphate (ATP). The importance of this stored energy becomes apparent in the relaxation of muscles. After death rigor sets in and no more ATP is produced.

The bulk fermentation of dough at 24°C for a period of approximately 2½ hours produces good volume and flavour. After this period the dough can be cut with a knife, remixed mechanically for a few minutes, and then left to ferment for a further 1½ hours. In the proving stage the still-fermenting dough is treated at about 36°C for just under an hour and attains its final shape and texture prior to baking. The baking temperature kills all the enzymes and changes the chemical nature of some of the products of fermentation to produce the final flavour of the finished loaf.

Fermented buns contain a certain percentage of enriching agents such as fat, sugar and eggs. This will tend to retard the fermentation process; the quantity of yeast used is therefore greater than that for a similar quantity of flour which is fermented for the same length of time.

The temperature at which a ferment is made affects the rate of its capacity to produce chemical changes. Yeast works increasingly faster at high temperatures than at a low one. At a low temperature the yeast takes a long time to produce gas, but while doing so retains the viability. A higher temperature ferment produces gas very rapidly but soon becomes exhausted.

Starter cultures are used in sausage emulsions to produce a lactic acid type of fermentation. These lactobacilli produce a tangy flavour. The organisms work best at a low pH and are relatively tolerant to salt and nitrite. Nitrate reducing bacteria, such as micrococci, produce adequate nitrite for good colour, but if excess nitrite is produced the sausage will be discoloured as a result of 'nitrite burn'.

The length of the fermentation period is just as important as the temperature. If a dough is baked before fermentation has progressed far enough, the resulting bread has thick cell walls and is decidedly less tender than bread fermented sufficiently. The volume is also smaller. However, if fermentation is carried too far, the gluten becomes too soft to retain all the expanding gas bubbles and coarse texture and thick cell walls result. With extreme fermentation the dough smells sour, and a very coarse, heavy, compact bread of small volume and sour taste is obtained.

CANNING

In the early nineteenth century, experiments were being conducted on the feasibility of packing meat, fruit or vegetables into containers for future use. However, the spoilage rate was high due primarily to incorrect sized containers, but also to the lack of penetration of heat to the food product. As a result the food temperature was below that required for sterilisation, and pathogenic bacteria multiplied to a dangerous level. Lack of knowledge, particularly of the scientific principles involved, must be regarded as the major cause of failure.

Two basic principles are involved: (*a*) packing the food into an impervious container, usually a bottle or steel container, lined with tin; and (*b*) applying a temperature high enough to kill all the natural micro-organisms present and thus produce sterility. This sterility will persist unless the container is opened, although certain products, as a result of internal molecular rearrangement, can produce compounds which may corrode the metal of the tin or produce drastic flavour alterations.

The production of a large number of cans of fruit or meat depends upon taking the product in its prime condition and subjecting it to an initial cleaning. Sterilisation will complete the process of microbial destruction, but a preliminary cleaning will do much to reduce the number of organisms involved. After cleaning, a blanching process follows where boiling water or live steam is used. Blanching will drive out any unwanted air bubbles which, if left, might prevent the attainment of a correct sterile temperature. Enzymes, which may affect the food stability, are also destroyed. The flavour and colour of peas is improved. In the filling process the cans are filled automatically and passed on to an air exhaust device where air and other gases are removed by high vacuum. Sealing the can and heat sterilisation complete the canning process.

Canned meat receives a heat treatment that will kill all vegetative bacteria and spores that might germinate and grow under normal un-refrigerated storage. However, an exception is the spores of thermophilic clostridia. It is essential to cool the cans rapidly after processing, and to avoid storage at temperatures above 35°C. The amount of heat required depends upon such factors as pH, presence of curing salts and the number of spores anticipated in the product. A semi-preserved meat like a canned ham is heat treated to kill all vegetative bacteria and produce good quality under adequate refrigeration. A common spoilage organism is a lactic acid bacterium which can produce discoloration and souring of the product after it has been removed from the can and sliced.

A number of food spoilage bacteria use carbohydrates as a source of their energy. Effective canning conditions will prevent this happening, and the problem is confined to either fresh meat or cured meat products. Meat products have better quality when they are sterilised using high temperatures and a quick process.

With tinned fruit where the pH is low (that is, below about 3.5), bacterial spores may survive the heating process; but since they cannot grow at this low pH no harm results. In non-acid foods such as fish and vegetables an autoclave is used to produce steam under pressure at a temperature of about 120°C, which will render the food safe after storage. As the temperature rises above 105°C there is a dramatic fall in the time taken to render harmless any micro-organisms present.

In conclusion it is safe to say that, apart from *Clostridium botulinium*, any micro-organisms that escape the sterilisation process will not be fatal. Nevertheless, any spoiled canned food should be discarded as any discoloration, odour or swelling is undesirable and can produce both physical and aesthetic discomfort.

INDUSTRIAL MICROBIOLOGY

The ability of micro-organisms to attack substrates and produce end products of a practical nature has been realised in many aspects of the food industry. In all cases the substrate may be regarded as a raw material, and the micro-organisms as the mechanism for changing the raw material into new products. Therefore the general reaction is:

Substrate + micro-organisms = end products (dissimilar or synthetic).

Provided the organism is readily available, and the raw material inexpensive, it is economically possible to convert this reaction into a large-scale industrial operation. Certain prerequisites are needed in industrial microbiological processes, particularly those involving fermentations.

Each organism selected must have the ability to produce large amounts of the product. The effect is in some ways catalytic. It should be able to grow rapidly and vigorously, be of a stable character and of course be non-pathogenic. The medium and the substrate from which the organism produces the new product should be cheap and available in large quantities.

It has been found possible to utilise the nutrient-containing waste from the dairy industry (whey) as a medium for culture purposes. The nutrients in the whey can be 'worked up' into products which form a vital part of animal feeds. After their formation these products may be present in a large volume, tank capacities of 200 m^3 being quite normal. It is left to the chemical engineer to design equipment which can isolate the required product from a heterogeneous mixture of microbial cells, unused medium constituents and other products of metabolism.

Moulds, yeasts and bacteria are all employed industrially. Even viruses, despite their small dimensions, have been used to produce vaccines. The wide variety of products produced include: (*a*) enzymes, (*b*) foods, (*c*) beverages, (*d*) antibiotics, (*e*) chemicals and (*f*) vitamins.

Processes and products from bacteria of interest to the food technologist are shown in Fig. 38. From whey, selected strains of lactobacilli are able to convert lactose into lactic acid. The whey represents a satisfactory medium since it contains, besides lactose, nitrogenous substances, vitamins and salts. Biochemically the organism produces lactic acid in a two stage process:

$$C_{12}H_{22}O_{11} + H_2O \xrightarrow{\text{enzyme lactase}} \underset{\text{glucose}}{C_6H_{12}O_6} + \underset{\text{galactose}}{C_6H_{12}O_6}$$

$$\underset{\text{glucose and galactose}}{C_6H_{12}O_6} \xrightarrow{\text{complex enzyme system}} \underset{\text{lactic acid}}{2CH_3CHOH\ COOH}$$

With vinegar, there are again two types of biochemical changes involved. Carbohydrate must be degraded to alcohol by a fermentation process, and then oxidised to acetic acid. There are several kinds of vinegars but the primary difference between them derives from the kind of material used in the alcoholic fermentation. The material selected would be fruit juices, hydrolysed and starchy materials or sugar-containing syrups. The end result should be a solution which contains not less than 4 per cent acetic acid. Oxygen is required as the process is aerobic.

Acetic acid bacteria, species of the genus *Acetobacter*, require constant aeration and work best between the temperatures 15 and 35°C. Any deviations from this range will not only inhibit the growth of *Acetobacter* but also permit the growth of other groups of micro-organisms that alter the desired biochemical change. Ultimate oxidation of the alcohol will produce acetic acid:

$$\underset{\text{ethyl alcohol}}{2CH_3CH_2OH} + 2O_2 = \underset{\text{acetic acid}}{2CH_3COOH} + 2H_2O.$$

The preparations of both lactic and acetic acid have involved the creation of substances of a simpler nature than the original substrate. However the biochemical process can also be synthetic and give a product more complex than the raw material. Dextran is a complex polysaccharide produced from the disaccharide molecule sucrose, the bacterium *Leuconostoc mesenteroides* being the selected organism. Complex chemical reactions are involved but the general reaction can be depicted as follows:

$$\text{Sucrose} \xrightarrow{\text{Enzymes from } L.\ mesenteroides} \text{Dextran} + \text{fructose.}$$

It is interesting to observe that although the dextran produced has a molecular weight of several millions the simple sugar fructose of molecular weight 180 (sucrose 342) is also formed.

Product	Micro-organism	Substrate	Culture type
Lactic acid (animal feed)	lactobacillus delbrueckii	corn starch or whey	anaerobic
Bacterial amylase (starch ferment)	bacillus subtilis	vegetable protein and sugar	surface aerobic
Dextran (stabiliser in food)	leuconostoc mesenteroides	sucrose plus nutrients	anaerobic
Sorbose (ascorbic acid)	acetobacter suboxydans	sorbitol plus yeast extract	aerobic
Cobalamin (food and feed supplement)	streptomyces olivaceus	dextrose	aerobic
Bacterial protease (meat tenderiser)	bacillus subtilis	protein, salts and carbohydrate	aerobic

Fig. 38.   Food products produced by bacteria.

WATER

In the last century it was proved that cholera epidemics had their origin in water. Since that time, an extensive study has been conducted into methods capable of detecting, identifying and controlling the organisms that concern public health and hygiene. A more pleasing aspect of water's microbial flora are those micro-organisms capable of producing chemical changes vital in maintaining the balance of marine life.

Lakes, ponds, rivers and streams constitute surface water and are prone to contamination from soil, air, domestic and industrial wastes. It is therefore impossible to generalise either on numbers or types of micro-organisms that can be found. Wastes can also provide nutrients or inhibitory chemicals to alter the microbial population. Therefore there is a vast difference between water from a mountain stream and that from a city harbour.

Other factors also affect the growth of micro-organisms, including:

(a) Temperature—water at 10°C would favour only psychrophiles and would represent an unfavourable environment for mesophiles and thermophiles.

(b) Salinity—halophilic types would predominate in sea or ocean water.

(c) Amount of dissolved oxygen—the growth factor of aerobes and anaerobes is related to the amount of oxygen in the water; some bacteria will live longer in water containing oxygen than they do in its absence.

(d) Presence of other marine organisms—algae provide a solid surface for bacteria and are also a source of food for them. Protozoa will utilise bacteria as a food supply.

Pathogens that cause infection in man are those that affect the intestinal tract. In addition to cholera there are dysentery and typhoid.

The organisms are present in faeces or urine from an infected person and may be discharged into a body of water that may ultimately be processed into a drinking water supply. Water analysis will reveal the presence of pathogens but it takes time.

Other organisms such as *Escherichia coli* are normal inhabitants of the large intestine of man and are in consequence present in his waste matter. These coliform organisms can be monitored and the presence of a large excess is a warning signal of pollution. The presence therefore of *E. coli* or other coliform organisms in water is clear evidence of faecal pollution of human or animal origin.

It is essential that water submitted for bacteriological analysis should be collected in a sterile bottle and be representative of the supply from which it is taken. There should be the minimum of delay between sampling and testing; if delay is unavoidable, the sample should be stored at a temperature between 0 and 10°C.

Routine tests would include (*a*) a plate count to reveal the size of the bacterial population, and (*b*) specialised tests to reveal the presence of coliforms.

The coliform test is spread over three distinct phases. The first consists of inoculating lactose-broth samples with appropriate amounts of water sample, incubating at 35°C and observing if gas is produced at 24 and 48 hours. The absence of gas after 48 hours proves that no coliforms are present, since this group of bacteria will produce gas from lactose. However if gas is produced it may be due to organisms other than coliforms.

Phase two therefore consists of transferring a sample from lactose-broth to a brilliant green lactose-bile broth (B.G.L.B.). This medium will inhibit the growth of lactose fermenters other than coliforms. Therefore if gas is produced in B.G.L.B. it constitutes a confirmed test, that is, coliforms are present.

Phase three would be aimed at identifying precise coliform organisms. By transferring a portion of the lactose-broth medium which has been inoculated into an eosin methylene blue agar plate (E.M.B.), the organisms will grow into recognisable colonies. *Escherichia* produce large dark colonies with almost black centres, and characterised by a greenish metallic sheen; *Aerobacter* will produce large pinkish colonies with dark centres and rarely show a metallic sheen.

MILK

As milk is a source of major nutrients, it is also an excellent medium for the growth of many bacteria. Milk contains an indigenous microbial flora from the cow, and it may be contaminated by further handling and processing.

Milk which is to be processed into milk powder either by roller drying or spray drying is accepted from the milk marketing board at a temperature of 7°C or less, and held in stainless steel tanks. It is customary to test the milk for its keeping quality by applying a methylene blue reduction test. The milk is added to a standard amount of methylene blue in a tube and then incubated in a water bath for 30 minutes. If the milk fails to decolorise the sample at the end of this period it is satisfactory.

Because of the large amount of milk handled at a processing plant the test may be abbreviated to last only 10 minutes but check controls are applied at a later stage. The bacteria in the milk use up all the oxygen dissolved in the milk, and then remove oxygen forming part of the methylene blue molecule. When the methylene blue is reduced it is decolorised. The methylene blue test is much more rapid than a plate count method and has more reproducible results.

A coliform test will be applied to milk, milk powder or cream on the lines previously discussed. In addition, where milk has been treated by the U.H.T. (ultra high temperature) process a colony count test is demanded. U.H.T. milk is treated at 132°C for 1 second. It can then be stored for periods up to 6 months at room temperature without adverse effects. In this test a loopful of milk is mixed with a yeastrel milk agar medium and incubated for 48 hours at 30–37°C. It is then examined for the presence of colonies. The test is satisfied if the number of colonies is found to be less than 10. Figure 39 shows the important effect that temperature has upon the numbers and types of bacteria present in milk samples.

Ropiness in milk shows itself by causing the texture of the milk to become viscous, resulting in 'ropy' or 'stringy' threads. Gum-like

Temperature (°C)	Number fluctuations	Bacteria present
1–4	slow decline initially, gradual increase after 7–10 days.	true psychrophiles e.g. pseudomonas.
4–10	Rapid increase in numbers after first few days. Large populations after 10 days.	as above, milk exhibits ropiness and proteolysis
10–20	very great increase in numbers. excessive populations in a few days.	mainly acid-producing such as lactic.
20–30	high density population within a few hours.	coliforms and other mesophilic types, gas and off flavours result.
30–37	as above.	coliforms favoured.
>37	as above.	some mesophiles but thermophiles favoured, e.g. bacillus coagulans, milk clumping.

Fig. 39.   Temperature effect on milk bacteria.

materials are synthesised by the micro-organisms and the viscosity increases. The composition of milk favours the synthesis of these materials. Proteolysis results in the production of soluble nitrogenous products that have an alkaline reaction, and give milk a bitter taste. *Pseudomonas fluorescens* will have a lipolytic effect upon milk fat; if the fatty acids produced have a low molecular weight the milk will have a sharp odour and flavour. Coliforms will produce 'unclean' flavours, and yeasts induce a characteristic yeasty aroma and flavour.

A number of dairy products including buttermilk, yoghourt and various cheeses owe their flavours to selected microbial cultures. Cultured buttermilk is produced by cultivating lactic acid bacteria in milk. The principal differences in production are related to the species of organisms which make up the starter culture, the temperature of incubation and the nature of the milk to be fermented.

Yoghourt has been popular in many European countries since the Russian bacteriologist E. Metchnikoff ascribed vigorous health and longevity to it. Milk is inoculated with *Lactobacillus bulgaricus* at 43 or 44°C. When the desired acidity, measured by pH determination, has been reached, the milk is cooled, and refrigerated overnight at 0–3°C. As the cessation of acid development is not instantaneous, it is necessary to start cooling from $\frac{1}{4}$ to 1 hour before the desired acidity is reached. If the yoghourt is cooled by a heat exchanger method this lag virtually disappears.

The presence of antibiotics in milk can affect the culture strains used. A concentration of penicillin as low as 0.1 μ/cm^3 may affect the growth of a yoghourt culture and 0.3 μ/cm^3 can stop it. The uncontrolled use of antibiotics for treatment of dairy cows for mastitis can produce difficulties for manufacturers of yoghourt cheese. A weak body and whey leakage in yoghourt can be traced to rough handling, wrong acidity and abnormal milk.

Cheese preparation involves adding a starter culture of lactic acid bacteria to fresh milk at an optimum temperature for curdling. *Streptococcus lactis* is used if the heat treatment does not exceed 38°C. If higher temperatures are used (approximately 50°C) *Streptococcus thermophilus* is selected. The enzyme rennin which is isolated from rennet extract obtained from calves' stomachs assists by coagulating casein. The majority of cheeses require ripening by bacteria or moulds after the curd is pressed into form. In the case of Roquefort, the curd is inoculated with the spores of *Penicillium roqueforti*, and then incubated at about 10°C in an area of high humidity.

The examples mentioned illustrate a variety of biochemical changes performed by micro-organisms during cheese ripening. A product of inferior quality, in terms of aroma or flavour, will result if conditions do not favour the selected organism. Dairy products are susceptible to spoilage (off flavours and aromas) by many types of 'undesirable' micro-organisms. Strict attention to detail is therefore very important.

FOODS

Due to the wide variety of metabolic activity shown by bacteria, yeasts and moulds, their effect upon food can be desirable or otherwise. Mould growth, although normally non-pathogenic, is certainly not welcomed on fruit but may be exploited to advantage in certain cheeses.

Accelerated freeze dried egg powder is pasteurised and should be free from pathogenic organisms but not necessarily free from spoilage organisms. The number of bacteria present in dried egg powder will depend on its contamination during handling and processing, etc. The viable count may range from a few hundred bacteria to several hundred millions of bacteria/gramme. Consequently any dilutions used from the original suspension should be noted so that a range can be obtained. In the laboratory a suspension of the sample in peptone water can be serially diluted and the samples cultivated in a dextrose tryptone agar medium for three days at 30°C. Both Gram-positive and Gram-negative bacteria will be found. Examples of Gram-negative bacteria include *Pseudomonas, Escherichia* and *Aerobacter,* and Gram-positive are represented by bacilli strains.

Micro-organisms can be involved in the spoilage of canned foods. Figure 40 indicates how the organism involved operates best over a certain pH range. As mentioned earlier, because of the heat-resistant nature of spore forming species of *Clostridium* and *Bacillus,* these constitute the most important group of micro-organisms in the canning industry. Other types of spoilage are more likely to affect quality rather than involve the consumer in dangerous illnesses.

Type and effect	pH range	Samples
Thermophilic bacteria		
flat sour	5.3 and above	peas
sulphide stain	5.3 and higher	corn
Mesophilic bacteria		
putrefaction	4.8 and higher	asparagus
rancidity	4.0 and higher	tomatoes
flat sour	4.2 and higher	tomato juice
Lactobacilli (poor flavour)	3.7–4.5	fruits
Yeasts (poor odour)	3.7 and lower	fruits
Moulds (surface growth)	3.7 and lower	fruits

Fig. 40.   Microbial flora in canned foods.

Pickles, sauerkraut and olives involve lactic acid bacteria for a desirable fermentation process. The micro-organisms present in the fermentation of cucumbers are in a brine medium. Temperature, salt concentration and the availability of fermentable substances control the type of fermentation and various groups of micro-organisms predominate at various stages of the process.

In the future, bacteria may be successfully harnessed to produce high-quality food from oil and natural gas. Research is going ahead into bacteria which can be 'farmed' and turned into food. It is envisaged that man would absorb the bacterial protein produced from chickens, turkeys, pigs and other animals reared on it. Micro-organic foods, which can contain up to 70 per cent protein, are certainly on the menu as the food of the future. These foods would have to be given body and colour to be acceptable but at the highest level they would be as nutritious and as tasty as steak.

EXERCISE 15

1. What types of bacteria are most likely to be present in spoiled canned foods?
2. Why is milk prone to attack by micro-organisms? Indicate any microbiological tests you feel necessary to control its quality.
3. What are coliforms, and why is *Escherichia coli* considered a pollution indicator?
4. Describe a laboratory method for water analysis. What treatment could be recommended to render it safe for use in food preparation?
5. Describe the characteristic features of industrial fermentation. What controls are needed for the quality aspects of a fermented food product?
6. Explain the terms
   (*a*) substrate, and
   (*b*) vegetative bacteria.
   Select two foods to illustrate your answer.

16 Food hygiene

IDEAL DISINFECTANT

Allied to disinfectants are antiseptics and germicides. At the simplest level, a disinfectant will destroy completely vegetative forms of bacteria which produce harmful effects in themselves or in the decomposition of matter. A germicide has the ability to destroy both useful and harmful bacteria, whereas an antiseptic may be regarded as a weak disinfectant. Concentration and chemical nature are the prevailing factors which separate the three groups. Phenol, in a concentrated solution, is a germicide, but at a much lower level is an antiseptic. Disinfection may be akin to chemical sterilisation.

It is to be expected that because of the many types of microbial cells encountered there is no single antimicrobial agent which is 'best' or 'ideal' under all conditions. An ideal disinfectant would possess a formidable array of specific characteristics. The chances are that no single compound possessing all these qualities will ever be found. A number of these features are described below and in the preparation of new and more powerful disinfectants these guide-lines should be followed.

1. The preparation should be homogeneous. Although a pure chemical is uniform throughout, a mixture of chemicals may lack homogeneity.
2. When using a disinfectant it should not be necessary to raise its temperature above that of the environment where it is to be used, in order to promote toxicity to micro-organisms.
3. Unless the selected compound or compounds have the ability to penetrate through surfaces their effectiveness is limited to the site of application.
4. The disinfectant should be non-toxic to both animals and man, whilst being extremely toxic to micro-organisms.
5. There should be no tendency for the disinfectant to combine with extraneous organic material. For example, some disinfectants have an affinity for protein. When such disinfectants are used where there is considerable organic material besides the bacterial cells, there will be little, if any, of the disinfectant available for action against the micro-organisms.
6. The disinfectant should be odourless or have an acceptable odour.
7. A disinfectant should possess some detergent activity so that the cleansing action contributes to the effectiveness of the disinfectant.

8. The disinfectant should be economical and available in large quantities.
9. The substance should have good water solubility in both hard water and tissue fluids.
10. There should be no loss of efficiency in stored conditions; germicidal stability is of paramount importance.

Certain compounds are available which fulfil a number of the above criteria. The factors to be considered in the selection of any individual member depend upon the nature of the material to be treated, general conditions and the type of micro-organism to be destroyed. The general conditions involve factors such as pH, time, concentration and temperature which must all be considered before application commences.

PHENOL AND PHENOLIC COMPOUNDS

Phenol, or carbolic acid, was studied extensively by Lister in the nineteenth century for its anti-therapeutic effect upon bacteria. The crystals are corrosive to the skin and care is needed in making up the solutions. It is still a popular disinfectant, despite the fact that others are available which are more effective in considerably higher dilutions. Phenol is a standard against which other compounds can be compared for their toxicity towards bacteria.

Solutions of 2–5 per cent are found to be effective as a disinfectant for working surfaces and food containers. Spores and viruses are more resistant to phenol than are vegetative bacteria. Low temperature and soap are detrimental to the germicidal property of phenol.

Phenol is a simple substituted benzene compound, and a number of derivatives of phenol have been prepared which are more effective than the parent substance. Figure 41 illustrates the chemical nature of phenol and its derivatives.

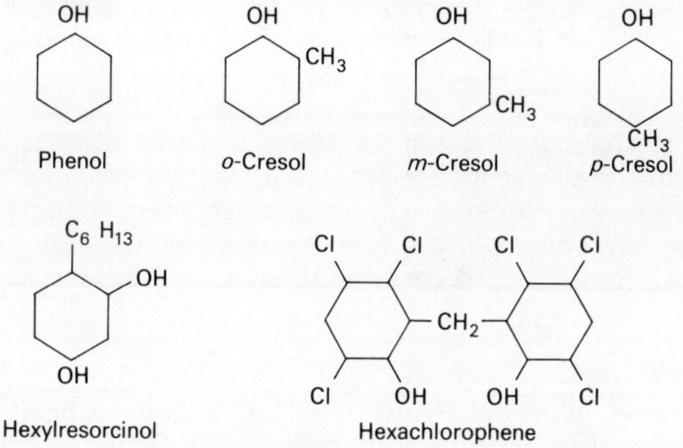

Fig. 41.   Phenol and Phenolic compounds.

The homologues of phenol (cresols) have approximately the same activity, which is several times greater than phenol. Tricresol represents a mixture of all three isomers. Cresols are not too soluble in water, but readily form emulsions with soap and alkalis. Hexylresorcinol is marketed in a solution of glycerine and water. It has a low surface tension and is therefore very effective against bacteria. General antiseptics are based on hexylresorcinol.

When chlorine enters a phenolic molecule a highly germicidal product results. Hexachlorophene, which has received television publicity, is in this category. It is insoluble in water but is readily soluble in alcohol and dilute alkalis. In very high dilutions it is bacteriostatic. Recent tests have shown that against *Staphylococcus aureus* it is effective in dilutions of 1 in 2 500 000. It is more effective against Gram-positive than Gram-negative organisms, and it can be implanted into soap. It is therefore ideally suited for personnel concerned with the packaging of perishable foods.

The mechanism governing the germicidal effect of phenol and its derivatives is believed to involve denaturing the cell protein and damage to the cell membrane. As a result the organism is altered to such an extent that it is destroyed.

ALCOHOLS

Ethyl alcohol is often used as a skin disinfectant; concentrations between 50 and 70 per cent being quite effective. It cannot be relied upon to produce a sterile condition, for concentrations which are potent against vegetative cells are practically inert against bacterial spores. Methyl alcohol is less effective against bacteria than is ethyl alcohol, and the fumes may produce permanent injury to the eyes. Higher alcohols such as propyl, butyl and amyl, are more germicidal than ethyl alcohol. There appears to be a progressive rise in germicidal power as the molecular weight of these alcohols increases. However, as the higher alcohols are not miscible in all proportions with water, they are not commonly used in disinfectants.

It has been suggested that alcohols owe their germicidal activity to the ability to produce coagulation in proteins. Their property of being effective dehydrating agents may be a disadvantage, because alcohols of 80 per cent strength will remove so much water from a cell that the alcohol cannot exert a penetrative action. Alcohol has a cleansing action on a surface, which will contribute to its ability to disinfectant.

IODINE

The pure element is only slightly soluble in water but is readily soluble in alcohol or aqueous solutions of potassium or sodium iodide. In this form it is a powerful germicidal agent, and is referred to as 'tincture of iodine'. The iodine concentration in these preparations is usually of the order of 5 per cent.

Iodine cannot be described as a selective bactericidal agent, because it is effective against all kinds of bacteria, and there is evidence of activity against spores. It is highly fungicidal and is to some extent virucidal. Because of its colour and odour it cannot be used on food surfaces, and it is chiefly used as a skin disinfectant. The effectiveness of iodine preparations (and alcohol) for the reduction of skin microbial population is shown in Fig. 42. Iodine preparations can be used as water disinfectants and for the sterilisation of food utensils.

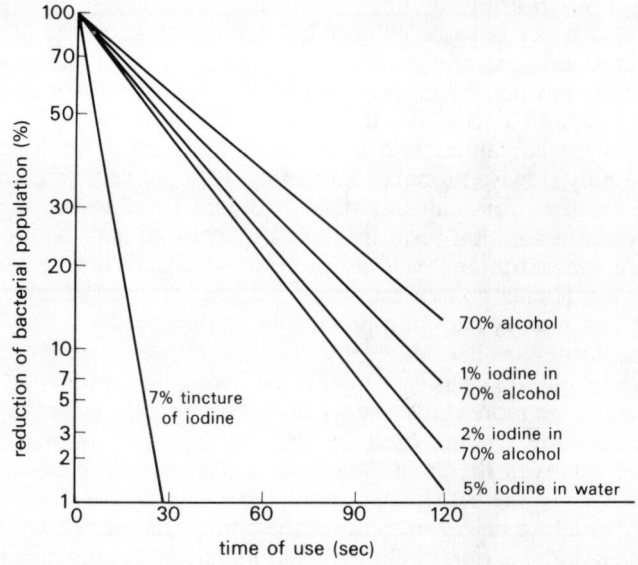

Fig. 42.   Iodine and alcohol as skin disinfectants.

No clear explanation has been put forward for the mechanism by which iodine exerts its anti-microbial activity. One theory is that the action might involve the combination of iodine with some cellular protein (perhaps enzyme), resulting in the destruction of a vital cell substance.

CHLORINE AND CHLORINE COMPOUNDS

Probably the most widely used germicides in food plant sanitation are those in which chlorine is the active ingredient. The compressed gas in liquid form is almost universally employed for the purification of water supplies. Due to its highly toxic nature it is difficult to handle unless special equipment is available for its dispensation. Canned meat products rely upon water treated with chlorine to neutralise any dangerous micro-organisms which might be present in untreated water.

Many compounds of chlorine are available which can be handled more conveniently than free chlorine and which, under controlled conditions of use, are equally effective as disinfectants. Factors such as pH, temperature and the amount of organic matter will affect the potency of chemical compounds of chlorine on killing bacteria and spores. They are most successful at a pH of 6 or less and at temperatures above 55°C. Any nitrogenous organic matter inhibits the germicidal action of chlorine. It is therefore essential that equipment should be thoroughly cleaned prior to to the use of sanitising solutions of chlorine compounds. In addition, if recirculated water is to be chlorinated it must be kept free of organic contamination.

Bleaching powder, or calcium hypochlorite, represents a convenient stored source of chlorine as a dry powder. Provided the powder is kept in an airtight container, and away from moisture, its efficiency will decrease only slowly. Products containing 5–70 per cent calcium hypochlorite are used for sanitising dairy equipment and eating utensils in restaurants. In solution form, sodium hypochlorite may be used. At a 1 per cent concentration it is ideal for personal hygiene or as a household disinfectant. Higher concentrations of 5–12 per cent are suitable for use as sanitisers in dairy and food-processing establishments.

Hypochlorites are the salts of a weak acid (hypochlorous) and cannot retain their chlorine easily. Organic compounds, represented by the chloramines, are more stable; hence they prolong the release of chlorine. Chloramines are characterised by the fact that one or more of the hydrogen atoms in the amino group of a compound are replaced with chlorine. One of the best known examples is Chloramine-T which has a complex structure based on toluene (the homologue of benzene).

The germicidal action of chlorine and its related compounds depends upon the production of unstable hypochlorous acid, which is formed when free chlorine is added to water:

$$Cl_2 + H_2O \longrightarrow HCl + HClO.$$

Both hypochlorites and chloramines are hydrolysed with the formation of hypochlorous acid. The hypochlorous acid in every case will decompose as follows:

$$HClO \longrightarrow HCl + O \text{ (nascent oxygen)}.$$

The exact way in which chlorine kills bacteria is a matter of some speculation. It is believed that the oxygen released in the above reaction acts as a strong oxidising agent, and through this action micro-organisms are destroyed. Chlorine probably combines with some cellular substances poisoning the cell.

Other bactericides which may be non-chlorine in nature, such as quaternary ammonium compounds, may be used as alternative sanitation chemicals at meat plants. They are based upon alcoholic solutions of ammonia and an alkyl halide. It may be difficult to remove chlorine from a food surface without excessive washing but this must be done properly otherwise the meat will be tainted.

SOAPS AND SYNTHETIC DETERGENTS

A detergent is a substance which acts with water to clean objects. It should be able to loosen debris (dirt, decayed food, etc.) and make it soluble or dispersible in water. The universal detergent, like the universal disinfectant has never been isolated.

Detergents tend to be specific in their action. Some of the desirable properties of a good detergent are:
1. No toxic effect upon animals or humans.
2. Capable of reducing the surface tension between the object to be cleansed and water.
3. Should possess some anti-bacterial properties.
4. No unpleasant flavour on the article to be cleaned.
5. Good emulsifying properties.
6. Good solubility in hard water.

Soap represents the earliest manufactured detergent. Basically, soap detergents are produced by a saponification process in which animal and/or vegetable oils and fats are heated with a strong alkali. Caustic soda will produce a hard soap and caustic potash a soft one. Animal fats used include mutton, tallow and beef whilst vegetable oils are represented by coconut, soya bean and groundnut. Chemically, soaps are the sodium or potassium salts of complex fatty acids. An example of a hard soap is sodium palmitate whilst a soft soap would have potassium instead of sodium while the organic portion would be derived more from liquid oils than solid fats.

The production of synthetic or soapless detergents is a relatively recent innovation. As living standards improved the drudgery associated with hand washing by soap had to be eliminated, particularly when automatic washers came on the scene. Soapless detergents are produced as washing powders, or as liquids, which will not affect the skin since they are not based upon strong alkalis. Both soap and a soapless washing powder perform a similar function, though the latter has quite a different composition. The bulk of synthetic detergents are based upon chemicals obtained from mineral oils and reacted with sulphuric acid (sulphonation).

Soap and organic synthetics both have a molecular structure which has two precise parts. They may be regarded as having a hydrophilic head and a hydrophobic tail. The hydrophobic portion will push between the water molecules and reduce the water's surface tension. Therefore the water collapses and increases in surface area. Any article to be cleaned is now 'wetted'. Grease is attracted to the hydrophobic tail and is squeezed away from the surface to be cleansed. The hydrophilic portion of a detergent molecule enables it to dissolve in water. Once the surface tension of water is lowered, the detergent molecule can begin to remove debris effectively.

Soap is capable of ionisation to produce metallic ions (positive) and non-metallic ions (negative). The non-metallic organic portion, for example stearate or palmitate, which is responsible for cleansing, would move towards the anode of an electrical circuit. Soap is therefore described as an anionic surfactant. Other surfactants might have an effective positive charge and would be attracted towards the cathode; such surfactants are therefore cationic in nature. Any surfactants which produced no charge in solution would be non-ionic.

The tails of soapless detergent molecules consist of long hydrocarbon chains, but the head portion has water-soluble properties and is sulphate or sulphonate in nature. They behave as anionic surfactants and the majority therefore resemble soap. Chemically, an anionic sulphonate would be produced by sulphonating alkylbenzene. Non-ionic surfactants are produced from long chain alcohols which are reacted with ethylene oxide. They are useful because of their high degree of solubility, and because of this they are often incorporated into liquid products.

Soapless detergents are able, because of their increased solubility, to lather more easily than soap. The presence of foam in a detergent is more a psychological attraction than physically beneficial. Foam is related to the use of tetrapropylene benzene as a raw material. The detergents which resulted could not be degraded by bacteria because of their complex molecular structure. Huge foam banks on rivers and settling ponds in sewage treatment works presented dangerous problems. Products have now been developed which have a simpler structure, and can easily be broken down. These detergents are known as soft or biodegradable and are the type normally found in Britain.

Grease and dirt represent debris on materials; because of their relatively large size they can be loosened by water and detergent. However, blood stains present a difficult problem, particularly if allowed to dry. Because of their colloidal nature they are very minute in size and resist attack by a purely physico-chemical process. By incorporating an enzyme which would biochemically degrade blood and other stains of a proteinaceous nature, it is now possible to remove them without damaging the fabric.

These enzyme detergents may cause a slight rash on the hands of some people with sensitive skin, rather like that of an acid irritant, although there is no chemical reaction. Research on detergents is still continuing despite the great advances already made.

The efficiency of a cleaning routine, especially in food preparation premises, can be determined by taking swabs from surfaces after the cleaning has been performed. An excellent swab made from 'Calgitex' wool can be used. The swab has a calcium alginate composition which will dissolve in weak Ringer's solution containing 1 per cent calgon (sodium hexametaphosphate). Ringer's solution is a physiological saline which contains the chlorides of sodium, potassium and calcium. It is widely used for sustaining cells or tissues during living biochemical experiments. A sterile swab is dipped into the calgon Ringer's solution and any excess squeezed out. The swab is then rubbed firmly over the surface to be examined, using parallel strokes while the swab is slowly rotated. The surface should then be swabbed a second time, using parallel strokes at right angles to the first set. It is necessary to see that the whole of the predetermined area is thoroughly swabbed. Place the swab in the tube of calgon Ringer's solution and agitate to dissolve the wool. At this point serial dilutions can be prepared if large numbers of bacteria are expected. A predetermined amount of suspension is then pipetted into labelled petri dishes containing suitable molten media (50°C). Mix in the approved manner, allow to cool and set, invert and incubate at 30°C for 3 days.

The colonies which have been produced can be estimated as the number of bacteria/cm^2 of surface and identified by Gram's staining method and specialised microbial techniques.

COMMERCIAL CLEANING AGENTS

These combine the properties of sterilisers and detergents. Five types are recognisable: iodophors, phenolic compounds, quaternary ammonium compounds, ampholytic detergents and hypochlorite and hypobromite compounds.

Iodophors always work in an acidic medium and are blended with synthetic detergents. They have a rapid broad spectrum of biocidal activity. The solutions can have low or high foaming properties and have a maximum operating temperature of about 50°C. Because of their brown colour they are easily seen. Their main use is with milk handling plants where the acid controls the formation of milkstone. The soft drink industry and breweries use iodophors for cleansing storage equipment by soaking or using low velocity circulation.

By contrast phenolic compounds are supplied as sterilisers only. Occasionally they may be blended as detergent sterilants and have excellent residual activity. They have an unpleasant odour but can be perfumed. Usually they form high foam liquids. Because of their taint they are rarely used in food vessels. Environmental hygiene is achieved with very small concentrations using manual procedures.

Quaternary ammonium compounds (alkyl substituted derivatives of ammonia) exist as sterilisers or detergent sterilisers and are marketed as liquids or in granular form. Compounds are generally of a neutral pH and produce foam in solution. They have good residual activity and are generally of neutral pH. Their biocidal activity is good but in a somewhat limited spectrum. They have a very wide use in the meat industry and in shop hygiene. However they are not recommended for beer use because the residue may affect beer head.

Ampholytic detergents are compounds with biocidal activity and can be blended with quaternary ammonium compounds and other detergents to broaden the lethal spectrum. They possess good residual action and are neutral liquids with high foam characteristics. As a group they are mainly involved in manual operations of a cleansing nature and in environmental hygiene. They have a limited use in cleaning and sterilising production vessels and are not generally suitable for spray or high speed circulation.

Hypochlorites and hypobromites are halogen-based inorganic compounds with pH adjustment for corrosion control. Formulations may have low alkalinity and high foam liquids for hand use, or high alkalinity and low foam for circulatory systems. They represent the broadest group of sterilisers available and are used in the food industry for manual, soak, spray and circulation cleaning. Application may be as a combined detergent steriliser where high and lighter residues are encountered or can follow a detergent clean as a terminal sterilisation.

IN-LINE CLEANING METHODS AND FLOW REQUIREMENTS

Grease is the main problem in the meat industry. In factories hot water with a strong detergent is effective if used at high pressure. Rinsing should be done after all washing operations using chlorinated water. For lairage or slaughter house, if steam is available, cleaning should be carried out with a steam lance using a medium strength alkaline cleaner at about 1 per cent strength followed by a good rinse with cold mains water. If steam is not available a medium strength alkaline cleaner (sodium metasilicate plus a detergent) at temperatures of the order of 70°C should be used.

The debris in scalding tanks will comprise grease and dirt from the intestines and fields. Wash out with cold water to remove particles and scrub with medium strength alkali (sodium metasilicate plus detergent mix) at about 0.2 per cent concentration. A temperature range of 45–55°C is preferred. Finally rinse with cold water. With stainless steel trays used for guts, etc. the previous conditions apply if washed by hand but if automatic washers are installed increase the concentration to 0.4 per cent and raise the temperature to 70°C.

Conveyors, tables and working surfaces in the casing room and meat preparation departments can be initially rinsed down with cold water. The debris may be a mixture of blood, grease and intestinal contents and will respond to treatment with mild alkaline detergents (0.2 per cent at 45–55°C) followed by a rinse with cold mains water. Floors and walls of brine rooms should be scrubbed as follows. Hose down with cold water and scrub with a long-handled, hard bristle brush using a solution of recommended detergent (alkali plus detergent mix) followed by a cold water rinse. A similar procedure may be adopted for sausage and small goods. Mincers, choppers and filling machines must first be scraped to remove residues and then dismantled followed by a swill with cold water and a detergent wash (alkali plus detergent base) at 45–55°C. A final rinse with cold mains water completes the cleansing operation.

The smoke racks in bacon curing factories require cleaning with 0.2 per cent sodium orthosilicate at 45–55°C. Allow to soak for 2–3 minutes and then use a scalding brush followed by a cold water rinse. With smoke chambers the interior may be similarly cleaned but if heavily contaminated, brush on a strong solution of sodium orthosilicate and leave for 1–2 hours. At the end of this period scrape, brush and rinse with cold water. Protective footwear in the form of wellington boots should be worn by operatives.

The dairy industry has to remove many kinds of matter, ranging from dried milk to protein scale. Acid detergents and quaternary ammonium compounds are efficient. Food utensils, vats and tanks may be cleaned by soaking in a detergent solution. All the surfaces must be pressurised to remove loose matter. Circulating water through the tanks also helps. With vats which contain heating coils the sequence would be to soak for 15–30 minutes, power brush, cold rinse, hot rinse at 85°C, drain and dry. Milk holding tanks are large vessels, and when cleaning them, spraying is best. Fogging type nozzles are used or 'tea drop' jets. The rate of flow is about 0.300 m^3/minute at a pressure of 3 atm. The sprayed liquid is at a temperature of 70°C and is water containing 0.5 per cent chlorinated detergent. It is important to use a spray pattern to cover all exposed areas.

Clean-in-place systems (C.I.P.) are especially popular in dairies as they possess the following advantages:
1. lower labour costs;
2. lower maintenance and repair costs;
3. cracks and crevices are eliminated;
4. shut downs due to leaks are avoided;
5. cleaning solutions can be recirculated after being brought up to strength; and
6. mechanical damage to pipelines resulting from repeated disassembly is avoided.

The procedure is as follows. Flush the line with water at 37°C until clean at the discharge end. Continue to flush out with an acid cleaner (20 per cent phosphoric acid, 3 per cent glycolic acid, 5 per cent wetting agent) at 60°C for 15 minutes. A final flush out with an alkaline cleaner (51 per cent metasilicate, 5 per cent sodium carbonate, 5 per cent wetting agent) at 70–80°C followed by hot water at 85°C for 5 minutes completes the cleaning process. The pipeline is drained and dried. Automated processes are now available from a central control panel but the cost is only justified on large units.

Recently, ultrasonic cleaning methods have been developed using frequencies of 30–40000 Hz. The ultrasonic energy is converted into mechanical energy with vibrations of the same frequency. The process is used in brewing especially in filter equipment. Immersion of the parts to be cleaned in detergent solution at 65°C is normal procedure.

EXERCISE 16

1. Why is it very unlikely that a single chemical agent will be found capable of controlling microbial populations under all conditions?
2. What examples are there in the food industry of chlorine and non-chlorine preparations as suitable germicides?
3. Discuss some of the attractive features of anionic detergents.
4. What problems of food hygiene may be relevant to high-class confectionery and how may they be overcome?
5. List some of the factors you feel influence the anti-microbial activity of chemical agents of the phenol category.
6. How would you ensure good hygiene in an abattoir?
7. What laboratory tests could be carried out to measure the relative ability of disinfectants against pathogens?
8. How would you as a food supervisor attempt to illustrate the importance of personal hygiene to workers on the shop floor?

17 Food spoilage

INSECT PESTS

The regulations that apply to the sale of food are numerous and are intended to protect the public from contracting one of many unpleasant diseases associated with poor hygiene. It is stated that any employee should 'take all steps as may be necessary to protect the food from dust, dirt, mud, filth, dirty water, animals, rodents, flies and other insects'. Despite all attempts to implement the regulations, a few cases will occur and these are often traceable to insects in particular.

There are many factors which account for food spoilage by insects. One is the vast numbers involved; locust plagues are a good example. In all insects there is a definite metamorphic cycle which comprises four stages: eggs → larva → pupa → imago (perfect insect). In spite of the hazards of predators and man's destructive ability enough survive to eat their way through thousands of tonnes of grain annually. When cereals represent a country's main source of food, such ravages can be disasterous. In a developed country the inconvenience may be financial as prices rise due to shortage.

A more serious aspect is the tonnage which is not eaten but contaminated with pathogenic bacteria carried on the feet of insects. The results may take several days to manifest themselves when a long incubation period is involved. Usually the eggs are either laid directly in the food product or in its immediate vicinity. Each egg will then hatch under suitable environmental conditions to produce a characteristic larva, e.g. maggot from the house-fly and caterpillar from the moth.

The larval stage is very destructive because most of the time is spent in eating prior to the dormant stage of the pupa. It may take a considerable time, depending upon the species, before the cycle is completed with the emergence of the perfect insect. With the common housefly the cycle is complete in less than a month and prolific breeding results in a vast population. Occasionally the cycle is incomplete as when the larva issuing from the egg is a replica in miniature of the adult; it is then known as a nymph. A cockroach is a good example of this modified metamorphosis.

True insects have three pairs of legs but mites which are not true insects have four pairs of legs. Mites not only contaminate food but can also produce skin irritation. Lack of moisture and poor ventilation are ideal for their propagation. When samples of raw materials, such as flour, are not stored under good conditions a visible insect population can be distinguished using a hand lens. It is also possible to detect parts of wing cases, pupal threads and larvae. As carriers of disease the housefly and the bluebottle are the chief cause for concern in Britain whilst mites give rise to food spoilage of raw materials.

CONTROL OF INFESTATION

There are two main methods of control available: physical or chemical. The method chosen will be dictated by the food supply, insect type and environmental conditions.

The use of an industrial type of vacuum cleaner is a popular physical method. It is of particular use in areas of heavy contamination due to a very specialised environment. Sacks containing raw materials should be swept, and the regular moving of sacks helps to disturb the routine of the insect cycle. The optimum range of insect life is between 5° and 34°C; above about 60°C all insect life perishes.

High temperatures are not advisable as a form of control, however, because of denaturation, but with some raw materials, particularly fats, cold storage is an obvious advantage. Adhesive bands similar to domestic fly-paper can be employed, and the use of ultrasonic waves is efficient, but expensive.

The physical principles of centrifugal force are employed in the construction of an entoleter, a device designed to eliminate insects by impact. A supply of food is placed on a revolving plate. Cohesive forces bind the food particles together but insects are thrown away from any nutrient and are killed or severely damaged by collision with a restraining surface. This is not particularly effective against insects with a hard outer chitinous cover.

Whenever the source of contamination is likely to persist, a food manufacturer will resort to chemical methods of pest control. This can be effective but, because of the danger of food contamination, requires considerable skill. All the chemical compounds used work either by (*a*) direct entry to the digestive system, (*b*) cuticle penetration, or (*c*) entry to the breathing system.

A number of poisonous compounds may be mixed with grain and spread in the path of insects; examples include sodium fluoride and boric acid. Sodium fluoride is very potent, but since it is toxic to animals and man particular care must be exercised.

The use of sprays against insect hordes represents an attack aimed at the cuticle. Absorption will occur and toxic compounds spread throughout the internal organs. Available chemicals include the pyrethins, gamma benzene hexachloride and dichloro-diphenyl-trichloroethane (DDT). These can be applied as dusts, sprays, smokes or emulsions. Pyrethins are extracts from the pyrethrum flower and are harmless to man and animals. They possess both contact and residual properties. Contact refers to their ability to attack either the cuticle or the breathing mechanism. Residual property means that the insecticide is potent for some time after application and can act whenever an insect walks over or touches it.

'Gammaxane' is another synthetic insecticide which is fatal to most insect types. It is both a contact and residual poison. Some brands may impart a musty odour to foodstuffs and care is required in its distribution, particularly in a bakery.

DDT is still an effective substance but may be a cumulative poison in man. There is also evidence of drastic mutation among certain species of pests whereby they develop strong resistance or immunity to it. Neither benzene hexachloride (BHC) nor DDT is of use on a lime-washed surface due to the hydrolytic effect of the lime.

Many commercial insecticides are a combination of the above types to meet the threat posed by a mixed population of pests. In order to produce an effective insecticide dispersion, many sprays contain a kerosene (light oil) base. Kerosene is highly inflammable and naked flames must be avoided.

Expansion of a successful business may be hampered by failure to obtain premises with a history of food use. Old warehouses which formerly housed cotton would be an example of areas where infestation may be extremely high, and against which normal spraying is ineffective. Gases are preferred because of their high powers of penetration. Fumigants such as ethylene oxide, hydrogen cyanide and methyl bromide are available. However, because of the high toxicity involved, the operation cannot be carried out until permission has been obtained from the local Medical Officer of Health and the Sanitary Inspector. The operation must also be carried out by experts and a permitted period of time allowed to elapse before the building is ready for machine installation, etc.

Food technology must not only work for the production of more food but also to preserve the food that is produced. Some authorities feel that the loss to food and industrial crops by all sorts of pests and diseases may account to 25 per cent of the original amount. Even if the figure is only 5 per cent it is one which the world cannot afford as populations increase.

BACTERIAL SPOILAGE

It is important to realise that food spoilage is not the same thing as food poisoning. Frequently one may lead to the other but this is not always the case. Again non-bacterial spoilage may resemble true infection as the following case illustrates.

A black coloration was observed in lacquered tinplate cans containing ham in jelly. It was noticed that discolorations on the surface of the contents did not occur where fat was touching the can. Discoloration was most prominent where the meat or jelly came into contact with certain points of the can seam. The lead content in the external portions of the meat was always higher at the discoloured areas. Analysis on the spectro-photometer revealed lead contents of 0.09–0.20 mg/kg of meat in the non-discoloured parts which contrasted strongly with the discoloured parts where the lead content varied between 105 and 1102 mg/kg of meat. The black coloration was caused by sulphur-containing groups in the meat protein reacting with lead from the soldering material used, black lead sulphide being produced. Because of the chemical reactions which can occur between meat pigments and bacterial metabolic products such as hydrogen sulphide, meat would assume a dark brown, almost black, coloration which closely resembles the previous case history.

Proteolytic micro-organisms are found in ham brines which contribute to the deterioration of salted ham during the summer. These organisms are bacteria of the *Proteus* variety. They do not grow, but neither do they die, in the high salt concentration. They may under suitable conditions produce poor texture and flavour.

In semi-preserved meats that are not adequately processed, lactic acid bacteria typified by *Streptococcus faecium* may survive. Discoloration of the product and souring will result, and are apparent after the meat has been removed from the can and sliced. Other surviving bacteria produce gas and can partially digest the meat and gelatin.

Pig's feet are often preserved in vinegar and the pH should be low enough to inhibit the growth of most bacteria. However, if any ferment-able sugar is present, and specialised acid-tolerant lactobacilli are in the medium, then turbidity appears.

Bacterial spoilage can still occur with semi-preserved meats which have been adequately processed but inadequately refrigerated. The appearance of gas, poor flavour and, ultimately, extensive putrefaction can be traced to the spores of bacilli and clostridia which have germinated and grown. Clostridia represents a variety of forms which are mainly associated with food spoilage but two species are responsible for true food poisoning by producing powerful toxins. *Clostridium botulinum* causes botulism, which is often fatal; *Clostridium welchii* causes a milder form of food poisoning, or it can cause gangrene if the organism enters the blood-stream or tissues.

Clostridia are often found in soil as saprophytes; a laboratory method is available for the cultivation and identification of various strains, most of which are proteolytic in nature. Soil and minced meat are mixed well in a beaker with a little water and the beaker is then placed in a poly-thene bag and incubated at room temperature. It is not advisable to incubate too long because of rife putrefaction. About 10 g of sample are removed, preferably in a fume cupboard, and transferred into a conical flask containing 100 cm³ of peptone water and glass beads. The flask is then placed in a water bath at 80°C for 20 minutes to destroy all the vegetative cells, and activate the viable spores which will germinate when the growth temperature range is reached on cooling. After cooling, the suspension can be allowed to inoculate tubes containing suitable media.

After incubation at room temperature for 24–48 hours the tubes can be examined for gas production, sulphide production, proteolysis, etc.

The genus *Clostridium* has many members which are vigorous fermenters; large gas volumes result and a wide variety of end-products are notable features. Acetic acid, acetone, butyl alcohol and butyric acid are representative of their degrading activities on food nutrients; canned meat may swell, or even explode, and the problem of offensive odours still remains

Bacon, if kept in close contact with sacking, may develop the phenomenon of 'sweating'. The air becomes fully saturated and the temperature of the bacon will rise. The resultant conditions are ideal for selected species of bacteria, and a translucent slime is produced on the bacon surface. Cultivation of this slime in suitable media reveals the presence of both Gram-positive and Gram-negative types in clusters of diplococci and staphylococci. Bacon slime only occurs in brine-cured bacon; normal dry salting would extract too much moisture for bacterial growth to occur.

Bacillus subtilis may produce spoilage in straw or grain, since it can be present in either the vegetative or spore condition. It is known as the hay or straw bacillus. Whilst the food source is in a dry condition, the spore stage predominates, but dampness can produce the free-living form with rapid growth. Some of the spores may be present in the final milled flour, causing damage to the flour prior to breadmaking. Fortunately this is rare, because control of the moisture content of flour, and modern storing methods, discourage spore germination. The addition of water to flour in dough making would produce some germination and it is possible that some spores in the interior of the loaf might not be destroyed. However, the chances of the formation of large amounts of growth sufficient to produce recognisable food spoilage at this stage is unlikely.

Occasionally, when bread is allowed to cool slowly or placed in warm surroundings, the condition known as 'rope' may manifest itself. When the bread is examined internally, a number of sticky threads are found between 1 and 3 days after baking. There may also be ester odours or fatty acid taint. The optimum temperature for rope development appears to be about 40°C. The cause is *Bacillus mesentericus* which may have been introduced into the bread through the yeast, flour, bread, bread box, mixing bowl or flour bin. The spores of this bacteria are not destroyed by the heat during the baking of the bread. Rope is very rare in practice, however, since chemical additives are available which make the dough more acid than normal (pH 5 or less) and this inhibits the bacillus. In plant bakeries the baked bread is rapidly cooled with circulatory air fans so that the temperature of 40°C is only experienced for a minimum period.

An even rarer occurrence of bacterial bread spoilage is 'bleeding bread' which shows itself as a red spot in the crumb. The bacterium responsible is *Bacillus prodigiosus*, which produces a red pigment as a result of its growth cycle. Water supplies and soil are the sources of origin. Modern techniques of chlorination ensure safe water for dough production, and soil infection is unlikely from employees.

FOOD POISONING

The number of reported cases, annually of food poisoning in this country is about 7000 but it is estimated that with non-notifiable cases the total may reach 100 000. On the Continent, British meat is only accepted if the gut is tied in evisceration to prevent the bowel contents contaminating the meat.

Staphylococci bacteria are found in the noses of 50 per cent of the population of any community. Bacteria are also carried by the skin which ejects microscopic particles into the air daily. An examination of culture plates containing bacteria shows that certain viruses can prey upon a bacterial load and hence reduce bacterial food poisoning. After incubation colour spots appear if the bacteria have been destroyed. By this technique both the source and patient may be cross linked.

The most common type of food poisoning in England, and widely distributed in a fatty food, is salmonellosis. It is a bacterial infection and, fortunately, large numbers of bacteria must be ingested before an infection results. The salmonella group comprises many species and one unpleasant feature is the ability of some individuals to act as 'carriers'. These people are not affected physiologically but can innocently infect others.

Meat products are occasionally involved but poultry products are more frequently the target for infection as the following example shows. About 150 people contracted salmonellosis after eating fried chicken. Investigation revealed that a member of the kitchen staff, a boy of sixteen, had the responsibility for cooking the chickens. He was given no adequate training and simply told to 'use his initiative'. The result was exterior browning without adequate internal heat and the organisms survived. A follow-up investigation revealed that the original supply of live chickens had salmonella by consuming grain which had not received any heat treatment. The manufacturers claimed that the involvement of heat would have added considerably to the retail price of the grain. This was not accepted and fines were accordingly imposed.

Staphylococcal poisoning is due to the production of a toxin caused by the organism growing. The symptoms are very similar to those of salmonellosis, with nausea, vomiting and diarrhoea rapidly manifesting themselves. Recovery is within three days and death is very rare, though some very elderly or very young victims succumb. A list of the foods most commonly involved in staphylococcal food poisoning would include cream-filled pastries, dairy products, poultry products and ham.

The most useful way of killing food poisoning bacteria is by the application of heat. All the non spore-forming organisms are destroyed by adequate cooking or pasteurisation treatment. *Salmonellae* and *Staphylococcus aureus* will be killed in stuffed poultry if the internal temperature reaches 75°C.

However, a complicating factor with staphylococci is that the toxin produced can remain active even after thirty minutes at 100°C. In this respect, spores of *Clostridium botulinum* need special care because of the heat resistant properties of the spores. Fortunately non-toxic *bacillus stearothermophilus* which could cause spoilage in canned uncured meats need longer heat treatment at a high temperature to destroy them. There is thus a large safety factor applicable to botulism.

In addition, the botulinum toxin is labile (relatively unstable) to heat and is destroyed at a temperature of 100°C. The rare cases of botulism are often with duck or salmon pastes which have not received adequate treatment, severe symptoms and even death being traced to active toxins. Figure 43 summarises the characteristics of the three established types of bacterial food poisoning.

Bacteria	Toxin	Development of symptoms
Botulism clostridium botulinum	unstable to heat	between 12–36 hr. occasionally 96 hr
Staphylococcal staphylococcus aureus	stable to heat	usually 1–3 hr, rarely 6 hr.
Salmonellosis salmonella (many species)	no toxin, true bacterial infection	usually 12–24 hr, occasionally up to 72 hr.

Fig. 43. Bacterial food poisoning.

A survey covering the three bacterial groups discussed above shows that over a period of years the most dangerous food sources of pathogenic bacteria have been manufactured meat products. Probably a major cause is the convenience factor of meat pies and considerable danger lies in the re-heating of these products. In the summer months Cornish pasties should be purchased only from premises with a high standard of food hygiene. Cream cakes, unless stored in cool conditions and away from flying insects, are another source of concern to the public on holiday. Vegetables should be well washed to remove soil particles and handled by people with clean hands. All pastry should be selected using plastic tongs. Employees with skin abrasions or sores should not be allowed on packing lines; infection here could be all too common. Figure 44 shows some of the widespread diseases that can be traced to salmonella organisms.

An occupational disease of farmers and butchers is brucellosis. In

Disease type	Natural host	Comments
Gastroenteritis (1)	swine and other animals	associated with hog cholera, passes to man via contaminated meat.
Gastroenteritis (2)	fowls	transmitted to man via dried eggs.
Gastroenteritis (3)	pigs and chickens	transmitted to man in eggs and meat.
Typhoid fever	man	transmitted by human carriers, flies, food and water.
Paratyphoid fever	man	similar to typhoid fever but with less severe symptoms, transmission features as for typhoid.

Fig. 44.   Salmonellosis in Man.

man the disease is not exclusively food-borne but occurs most commonly as a result of drinking milk or eating milk products that have come from infected animals. The species of bacteria producing bovine brucellosis is *Brucella abortus*. Infection in man is a direct result of organisms from infected animals entering the handlers through cuts or skin abrasions. After a long incubation period of up to 30 or more days, the victim complains of pains in the muscles and joints with irregular fever which continues to a chronic stage. The disease can be effectively prevented by using pasteurised dairy products; the danger to the general public is thus minimal. Calves may be vaccinated with weakened strains of *Brucella abortus* to reduce the incidence of the disease in cattle.

Food poisoning due to micro-organisms other than bacteria is minor. At this stage reference can be made to toxicity which can be traced to mould. A number of apparently inexplicable turkey deaths were reported on a farm. The symptoms appeared to be loss of appetite, and limb rigour followed by irreversible coma. An examination of the water supply for metal toxicity revealed nothing but a microscopic examination of the food (groundnut meal) revealed contamination with the mould *Aspergillus flavus*. Normally, moulds may be regarded as agents of food spoilage rather than pathogens. Unfortunately *Aspergillus* yields, as a product of its growth, aflatoxin which resembles the liquid toxin from *Clostridium botulinum*. Enough had been produced to kill animals the size of a turkey. The moisture content of the meal was over 16 per cent and this favoured mould growth.

The number of food poisoning cases which may be regarded as 'unpreventable accidents' can be counted on the fingers of one hand. It is the responsibility of everyone connected directly or indirectly with food to observe the highest standards of hygiene. Individuals may be powerless to halt, for example, atomic reactor pollution, but the control of bacterial pathogens is within the ability of every responsibly minded citizen.

Research is also available in the fight against infection. A recent investigation into botulism in canned cured meat in the U.S.A. reveals that their high safety record may be due to the use of nitrite in their processing. Tests on cured ham showed conclusively that sodium nitrite has a definite inhibitory effect upon *Clostridium botulinum* in perishable canned meats. The key factor appears to be the level of nitrite at the time of product manufacture, rather than residual nitrite concentration resulting from bacterial nitrite degradation. As a result of these tests a bacon study is under way and a frankfurter test is being developed.

EXERCISE 17

1. Which insect features render them a serious threat to increased food production? In your opinion what would be the best course of action to take to eliminate them from a general catering area?
2. Define botulism. Describe its causative agent and how it may be controlled.
3. To what extent may bacterial food spoilage be a problem in flour confectionery?
4. In your opinion why should meat and meat products be particularly vulnerable to pathogenic bacteria?
5. Illustrate, by reference to baking products, any visible signs of food spoilage. Discuss briefly the reason behind a selected contaminant.
6. Describe how you would plan a publicity campaign aimed at bringing home to your employees measures to control food poisoning in confectionery products.
7. Write brief notes on:
   (*a*) staphylococcal poisoning, and
   (*b*) salmonella poisoning.
   Your answer should include possible food sources of infection.
8. Comment on the statement that 'heat will render all foods safe for human consumption'.

18 Food preservation

INTRODUCTION

Most foods contain natural micro-organisms. It is only when many organisms have developed that food spoilage is evident, and when certain types have evolved that food poisoning is possible. In the preparation of certain cheeses and yoghourts the reverse process occurs, with micro-organisms producing a less perishable product from milk. Micro-organisms resemble higher organisms in that their metabolic activities can be arrested by depriving them of certain nutrients. Water is essential for their survival in some form or other. Heat and cold will destroy them or render them harmless. In the food field one can observe the versatility of micro-organisms. Some are undesirable and their growth must be controlled; others produce desirable changes of a preservative effect and they are exploited for this reason.

Food can be preserved by a variety of methods. Some are ancient, and depend upon prevailing climatic conditions for their success. The cold climate in Europe, thousands of years ago, acted as a natural preservative for prehistoric animals, as illustrated by the discovery of mammoths with edible flesh. Antarctica today represents an area in which micro-organisms are virtually absent. Any extreme species capable of surviving the cold conditions would not reproduce, and therefore pose no threat to food in a fresh condition.

Primitive man also learnt the value of preserving food by depriving it of water. The drying of fruit, vegetables and meat increased his chances of survival. Beer fermentation was practiced by the Egyptians, the alcohol acting as a destructor of food spoilage bacteria. Caves are always a source of low temperatures, even in the heat of Africa, and food placed in them will keep longer before decay renders it inedible.

The United Kingdom is the largest importer of meat in the world; for example, more than a quarter of a million tonnes of Danish bacon are imported annually. Legislation, such as the Carcasses and Animal Products Order of 1954, prohibits the entry of animal products from countries where animal disease has not been brought under control.

As recently as 1948 it was demonstrated that the virus of foot and mouth disease could exist in the frozen state in bone marrow for several months. In the epidemic of foot and mouth disease in 1967/8 the origin of infection was thought to be South American lamb; in consequence, lamb is not now imported from most of the South American continent. A ban has also been introduced on the importation of bovine offal and bone in beef.

A typhoid fever outbreak in 1963 was traced to a particular brand of corned beef which had been produced in an Argentine factory where the water used for cooling cans was untreated river water. Most probably the typhoid bacillus gained entry to a tin of beef after it had been sterilised. River water which was unchlorinated and tepid would contain vast numbers of typhoid bacillus (*Salmonella typhi*). It is accepted that any possible infection in a can before the sterilisation process will be eliminated because of the length of time during which the meat is subjected to very high temperatures. However, in subjecting cans to very high temperatures under considerable pressure for a period of hours, a strain must be imposed on the can's construction. Micro-leakage may therefore occur through defects in the tinplate or through faulty seams. Polluted water used in the cooling process transfers the organism to a suitable environment and no detection will be possible during the incubation period as *Salmonella typhi* will not produce gas.

All water used in meat plants overseas is required to conform to a minimum bacteriological standard of purity. This standard insists that the water must show no coliform bacteria in 100 cm^3 samples. The standard is maintained by close supervision, and regular and frequent sampling of incoming water supplies.

The introduction in the last decade of large containers for holding food, as an alternative to the conventional method of storing carcasses in large cold rooms, has posed a food preservation problem for Port Health authorities. It is not easily possible to examine individual items, and the container may not be unpacked until it has reached an inland destination. Containers speed the flow of imported food, but their use demands close co-operation between port and inland authorities to ensure proper inspection.

As there is no restriction on unloading a container before public health clearance has been given, a container may be unloaded whenever it reaches its destination. It is appreciated that, if it were considered necessary to examine all containers in ports, then the advantages of containerisation would be lost. Nevertheless, the effective examination of fresh meat transported in containers requires the provision of special facilities to prevent deterioration of the product and not all ports have these facilities or have access to them.

All methods of food preservation are based upon a number of basic scientific principles. First, it is essential to remove a contamination, or if possible prevent its entry. Secondly, methods should be sought to inhibit the microbial growth and prevent metabolism which might produce toxins. Thirdly, the destruction of any organisms which might produce food spoilage or food poisoning should be a major objective. We shall now examine the methods available.

PHYSICAL AND CHEMICAL METHODS

Industrial firms employ a large variety of techniques to preserve food, so that when it is returned to normal temperature the quality is very close to that of the product in its original natural fresh condition. Figure 45 summarises the types of modern general preservative methods available.

Method	Technical details
Aseptic handling	filtration, protective clothing, personal hygiene.
Heat	boiling in suitable containers; use of steam under pressure; pasturisation.
Cold	use of low temperatures; specialised refrigerants; freezing techniques.
Desiccation	water removal by specialised physical techniques and minimum denaturation.
Osmotic pressure	use of concentrated brine or syrup solutions with flavour control.
Chemicals	addition of chemicals direct. Smoking methods to produce preservations. Use of microbial fermentation products, e.g. acids.
Radiation	ultraviolet and ionising radiations.

Fig. 45.   Food preservation methods.

Heat is used to produce either a pasteurising or a sterilising effect upon certain food products. The effect is to destroy the organisms, and to preserve food packed in cans, jars or other types of packages that resist the entrance of micro-organisms after processing. Commercial food canning uses steam under pressure (autoclaving), a most effective method as it can be relied upon to kill both vegetative and spore forms of bacteria. It is ideal for non-acid foods, for example, meats and vegetables. The process is subsequent to elaborate cleaning of the food and commences with a blanching stage which removes excess gas and kills enzymes.

After peeling and coring, the food is transferred to its container and, after exhausting and sealing, is heated in pressure cookers at about 115°C for long periods (up to 1.5 hours/450 g), or it may be treated for shorter periods using higher pressure at 122°C.

The success of food preservation by heat is due to intensive research into knowledge of the heat resistance of micro-organisms, particularly spores. In addition, the rate of heat penetration through foods of different consistencies, and the size of the containers into which they are packed, have to be considered. Experimental techniques are carried out to determine the thermal death point of bacteria. For thermal death time (T.D.T.) determinations, suspensions of bacteria or spores are added to foods or media contained in the T.D.T. cans. After vacuum sealing, the cans are heated to various temperatures and for various lengths of time in small sterilisers. After this treatment the cans are incubated and subsequently examined for evidence of surviving organisms. Spoilage of canned goods is quite rare.

The hydrostatic steam steriliser will process thousands of cans per hour. Cans are drawn into the unit on horizontal grooves into which they are fitted in neat rows. The steam gives up its latent heat of vaporisation to the cans, and the contents of the can will heat by convection, if liquid is present (soups), or by conduction for mainly solid products (meats).

Canning is an ideal process from the viewpoint of storage after processing, because no special conditions are required. Its limitations include increased haulaged costs due to the metal weight, and applicability to foods which are eaten in a cooked form. Flavour deterioration in meat is possible after about 4 years, but a shelf life of 8 years is not unusual. With tinned fruit, the low pH can cause corrosion problems. Pineapples are below standard in 4 years, whilst red fruits deteriorate in under 2 years.

Canning has been extended to soft drinks and beers; the latter need special care because the flavour is very sensitive to contact with metal. In general, research is being aimed at heating for shorter periods but at higher temperatures so that the effect on palatability will be at a minimum. Milk has been treated at 140°C for 1–2 seconds to produce a sterilised sample without greatly affecting its taste or colour.

Pasteurisation is a milder form of heat treatment than canning. It will reduce the bacteriological load but as this treatment does not kill all the micro-organisms, it may be necessary to store the products at low temperatures. Milk and other foods such as fruit juices, are pasteurised. A disadvantage of pasteurisation is the survival of spores. Palatability, however, is not impaired and the nutritional value is higher than that of a comparable canned product.

The use of low temperatures as a means of preservation has increased enormously over the last 25 years. Temperatures approaching 0°C or below impose a static condition on the growth of micro-organisms. Modern refrigerants, and the use of new cryogenic materials, have made it possible to transport and store perishable foods for long periods of time. Some foods, for example vegetables, are scalded before freezing to kill microbes and enzymes. For long storage a temperature of −18°C or below is essential. Foods rich in fats do not store well; fish needs constant checking for deterioration.

Quick freezing techniques whereby food is frozen as rapidly as possible below a temperature of −18°C to minimise ice crystal formation has only reached commercial value since the end of the Second World War. As early as the 1920s the American, Clarence Birdseye, worked on the problem but the results were disappointing. Practical food engineering techniques had not been evolved, and the importance of maximum food freshness as the crop was harvested was not fully realised. Nowadays the range of quick frozen food is almost unlimited, including fish, fruits, vegetables, meats, confectionery products and complete meals.

Currently the main categories of quick freezing include air blast, plate and immersion techniques. Within these, a number of different techniques are available, each with advantages and disadvantages. The nature of the food to be processed and the sales potential of the finished product, will dictate to a manufacturer his choice of method.

Blast freezing involves the rapid movement of cold air at $-40°C$ over the product; the freezers in this category are discussed below.

1. Batch-type trolley freezers

These comprise an insulated cabinet which is manually loaded with food on trolleys. Refrigerated air is blown across the trolleys for a predetermined period, after which they are removed and the cycle recommences. The manual labour involved limits these freezers to small-scale operations involving meals for a specialised section of the public. Examples include prepared meals for use in hospitals and schools.

2. Fluidised freezers

The product is supported by jets of refrigerated air and an open wire mesh belt then conveys it through the freezing zone. Vegetables such as beans, peas and chipped potatoes, respond well. Soft fruits can be treated, but need more care because of their delicate texture. The product bed varies between 5 and 40 cm in depth, and the constant aerial agitation prevents the fruit or vegetables sticking to each other during freezing. Figure 46 shows a cross-section of a fluidised bed freezer.

Because of the high surplus water content often associated with vegetables, special attention is paid to the design of the refrigeration coils in order to prevent frost from rapidly affecting the transfer of heat from the air to the coils. Many systems are arranged to enable the coils to defrost continuously in sequence.

Fluidised bed freezers can have capacities of up to 12 tonnes/hour. Vegetables and soft fruits achieve complete freezing between 4 and 16 minutes.

3. Spiral tunnel

This is designed for limited space work. The conveyor belt carrying the food is wound round a rotating drum and conveyor lengths of up to 620 metres can be used in a single tunnel. This has an obvious advantage over a straight-line conveyor system. It is ideal for freezing thin products in 10 minutes. Refrigerated air can be blown across each layer of the spiral, enabling this system to obtain effective coverage of the belt. Thicker

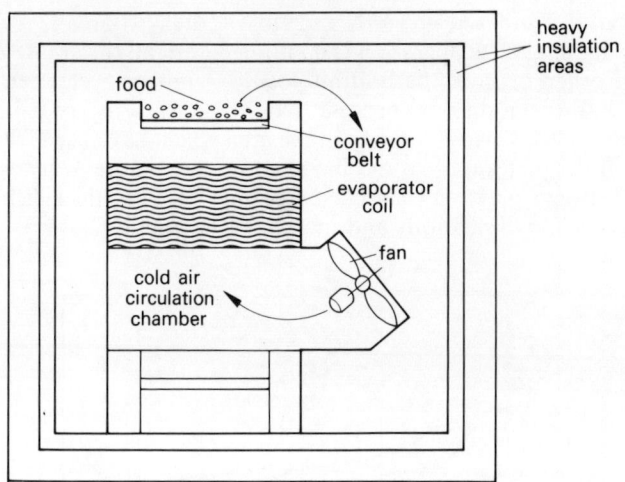

Fig. 46. Fluidised bed freezer.

products can be accommodated up to a freezing time of 3 hours. The range of products includes wrapped ice cream, fish fillets, beefburgers and meat pies.

4. Multi-tier tunnels

These are most useful for freezing thick wrapped food products requiring a residence time in the tunnel between 2 and 4 hours. The packages are loaded on to trays and the trays are pushed through the tunnel, the packages discharge at a lower level at the same end as the infeed. Heat transfer from the wrapped product will depend upon its thickness and the type of wrapping; freezing time will not be materially reduced by using a high air velocity. These tunnels are ideal for freezing poultry in boxes and other wrapped products contained in cartons.

Plate freezers are especially suitable for the freezing of food in block form since the use of pressure ensures the formation of high density blocks. Two types are found, known as batch type and in-line type.

In the batch type the refrigerant is evaporated inside aluminium plates, which are brought into direct contact with the food by contraction, and pressure is applied. The plates can be arranged in a horizontal or vertical configuration with two plates forming a freezing area. Products up to approximately 12.5 cm can be placed in the gap between the two plates. This freezer type is widely used in the meat and fish industries. A vertical freezer is often used for freezing liquid egg purées and fruit juices, by filling the liquids into plastic bags situated between the plates.

The in-line type is used in an automatic freezer system where cartons are loaded between the freezer plates held in the open position. When the plate is full it is indexed upwards and closed, the next area is then made ready for loading. The operation is continued until all the areas are filled. The plates then drop to the bottom position and the cycle then recommences, with the additional process that, as the cartons are fed into the loading areas, the cartons already in the area which have been frozen, are pushed off the plates on to a discharge conveyor. Figure 47 gives a comparison between plate and air blast freezing for 5 cm thick fish in cartons of the same dimensions and material construction.

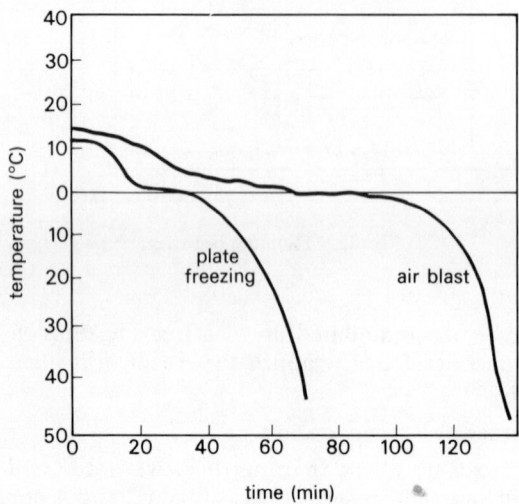

Fig. 47.   Plate and air blast freezing.

Immersion freezers can include non-toxic refrigerants in liquid form as represented by nitrogen, carbon dioxide and Freon. Foods are immersed in a freezing liquid or spray which freezes the food rapidly. The method is useful for the continuous production of meats and various sea foods. A note of caution has been sounded by the Food Additives and Contaminants Committee with regard to Freon as a direct contact refrigerant. Arcton 12 and Freon 12 both use dichlorodifluoromethane and the Committee are perturbed about the carry-over into food of organo-halogen residues. These residues are said to be very variable, as they can be high at the end of the freezing process but decrease fairly rapidly on storage. It is possible, according to the Committee's report, for properly thawed food, as consumed, to contain over 100 mg/kg of organo-halogen residues. Food which may be eaten when only partly thawed might contain much higher levels. Residues in cooked foods are generally lower, although in some cases they can still be upwards of 100 mg/kg, according to the report.

Freon manufacturers claim that the freezant possesses a short freezing time with negligible dehydration and is more economical than nitrogen or carbon dioxide since the freezant can be re-cycled with low loss. The Committee, although accepting that in isolated cases Freon has an economic advantage, believes that with conventional methods of freezing this will not be realised in practice. However the Committee is prepared to consider Freon as a direct immersion refrigerant on a product-by-product basis, and might in certain cases give consent for its use. Nitrogen and carbon dioxide, particularly the former, are recommended as immersion freezants as they have no residue problems.

With liquid nitrogen, the product to be frozen is placed onto a variable-speed stainless steel conveyor belt, which passes into an insulated tunnel where it immediately comes into contact with the counter-current flow of the cold gaseous nitrogen, the temperature of which progressively falls to $-196°C$ directly under the liquid spray, leaving the highest rate of heat transfer to take place in the small area under the spray, where half the freezing potential from the liquid nitrogen is utilised.

The basic advantage claimed for this faster, colder, freezing method is decreased damage to foods, which would otherwise be caused by the formation of large ice crystals. This is unavoidable in conventional, less rapid freezing techniques. Damage of this type causes high drip loss on thawing, with the result that valuable flavour, colour, and aroma constituents drain away. Dehydration losses in conventional freezing can be as high as 5 per cent, and with a high moisture content product this can seriously affect profit margins. In the nitrogen process, however, dehydration losses are estimated to be about 1 per cent. This is largely the result of the very small quantities of liquid nitrogen necessary to deep-freeze products. On vaporisation from a liquid at $-196°C$ to a gas at the same temperature each kg of nitrogen absorbs 47.7 kcal. Each kg of gas absorbs another 0.25 kcal for every °C it is warmed above $-196°C$.

A number of smaller meat firms have expanded their business as a result of using liquid nitrogen. Both uncooked pastry and processed meat products have been frozen on the liquid nitrogen freezer. Cornish pasties, steak and kidney pies, large and small sausage rolls, beef steaks, beef joints, lamb and pork have all been frozen on a production line basis. In one instance, the production rate for sausage rolls was increased fourfold with the use of liquid nitrogen.

Poultry farmers concentrating on chicken production have in some instances switched from conventional blast freezing to nitrogen spraying. They claim increased output; after treatment, the chicken portions emerge with a thin, complete layer of rock hard ice. This layer is designed to act as a barrier to external bacteria and unwanted heat.

Fresh fruit, particularly raspberries, have been successfully treated, one firm processing 300 tonnes of raspberries with liquid nitrogen in 5 weeks.

Scampi represents a perishable and costly shellfish product. One firm is freezing scampi successfully at a rate of 350 kg/hour, contrasted with 80 kg with the conventional freezing system it replaced. In operation, liquid nitrogen machines are simple to operate, and can therefore be run without a fully qualified refrigeration engineer.

Carbon dioxide as a liquid refrigerant is not as popular as nitrogen. In some cases it will convey increased acidity which, although slight, may affect flavour. Less research has been carried out on the quality of food frozen in this way. With carbon dioxide freezers liquid CO_2 is sprayed through nozzles into the tunnel at approximately $-78°C$. The design of the system is such that it allows the small particles of solid CO_2 to sublimate into the surrounding CO_2 vapour stream and this vapour is circulated by fans across the product at a temperature of approximately $-62°C$. Less loss on storage of liquid CO_2 will occur, compared with liquid nitrogen, and it becomes economic to install a CO_2 recovery system when freezing large quantities. This system is claimed to recover 80 per cent of the CO_2. The introduction of the recovery system, however, makes its capital cost similar to that of the conventional air blast freezer.

The selection of a particular freezer will depend upon whether the foodstuff is to be individually quick frozen or frozen in block form, and whether or not it is necessary to freeze in order to wrap. These factors, coupled with others such as running costs, physical size, desired quality and the control of density often prevent the selection from becoming a simple task.

Preservation by desiccation is more common than refrigeration in the food industry. The essence of the process is to remove water without materially altering the other constituents of the product. Growth and multiplication of micro-organisms are prevented by the lack of moisture, and also by the resulting increase in osmotic pressure. Fruits are commonly preserved by this method, common examples including currents, prunes and raisins.

In areas where fresh meats are not readily available and refrigeration is at a premium, dehydrated animal products are widely used. Dried milk and eggs are used extensively for domestic and commercial baking. If properly stored and protected from contamination, such products retain their nutritive qualities for long periods.

An obvious advantage of dehydrated foods is the great reduction in weight and bulk due to the removal of water. Hay, fodder and dried grains are examples of the use of desiccation for the preservation of animal feeds.

The preservative effect of desiccation is due mainly to a microbiostatic condition caused by the lack of moisture; the micro-organisms are not necessarily killed. Indeed this is dramatically demonstrated in the preservation of cultures by rapid drying in a frozen state. This technique is known as lyophilisation and many bacterial species can remain viable for 20 years. Essentially, the cells are frozen as suspensions using a mixture of solid carbon dioxide and alcohol. Upon the completion of drying the vials containing the samples are sealed under high vacuum conditions.

Bacterial cells placed in solutions containing large amounts of dissolved substances undergo plasmolysis. Water is lost from the cell and metabolism is arrested. This is equivalent to preservation by increased osmotic pressure, and is related in principle to inhibition by desiccation. Moulds and yeasts are relatively resistant to osmotic changes, but processes of food preservation based on this principle serve a useful purpose. Jams and jellies, because of their high sugar content, are rarely affected by bacterial action. The lactose and sucrose content of condensed milk assists in its preservation against microbial activity. A similar effect is observed by 'curing' meats and other foods in brines composed of high concentrations of salts.

Preservation by radiation in which bacteria are destroyed appears to be a promising new method for sterilisation of food without heat. It is sometimes referred to as a cold sterilisation process, as opposed to the use of heat which has been the basic method for sterilisation of foods for nearly two centuries.

Ultra-violet radiation has been used to lower the surface contamination of some foods. Meat processing plants can be equipped with germicidal lamps placed in the cold storage rooms to hold the microbial population within the area to a minimum. Longer periods of storage are therefore available without microbial spoilage.

Gamma rays, and high-energy electron beams, are being used experimentally as a means of preserving and extending the shelf life of certain foodstuffs. Their use introduces two problems: the development of adequate gamma-radiation sources for commercial practical applications, and the construction of equipment for safe operation. Although it has been demonstrated that pasteurisation and sterilisation can be achieved by radiation, much more research and development work must be done to arrive at a safe and economical processing procedure.

Chemical preservation has a limited application because of the possible toxicity of the 'safe' food when consumed. Only a few chemicals are legally acceptable for food preservation. These include sulphur dioxide, and certain salts containing sulphur (sulphites and metabisulphites, benzoic acid and benzoates, propionic acid and propionates.

Benzoate is used in ketchup and other vegetable products that are not heat processed, or are subjected only to mild heat-treatment of the order of pasteurisation.

Sulphur dioxide in water (sulphurous acid) is added to fruit pulp that is used in jam making and is almost entirely removed in the process of jam boiling. Acetic acid is a popular preservative for pickled sausages and pigsfeet; and propionic acid, usually as the sodium salt, is effective in dairy and bakery foods, especially bread. Moulds and other highly aerobic micro-organisms are inhibited by propionic acid. Sorbic acid is used as a fungistatic agent in such foods as citrus products, cheeses, salads, syrups, candies, and margarine to limit the growth of moulds and yeasts. Sodium metabisulphite represents a convenient source of sulphur dioxide in the preservation of sausages.

Although considerable emphasis has been placed on freezer drying as a means of preservation it is true to say that dehydration by heat is still carried out on a large range of foods. The merits of roller, spray and tunnel drying techniques can now be briefly considered.

Roller drying is a relatively simple economic process and is used by large companies for producing milk powder for a variety of purposes. In principle milk is allowed to fall on to the surface of a large heated roller; steam is the heating agent. Rapid evaporation occurs almost immediately and the milk collects as a thin film which can be manually or automatically removed by a thin steel blade set close to the plane of the roller. The process demands a film thin enough to be dried at a temperature not too high for extensive denaturation to occur. However, it is inevitable that some 'cooking' occurs with a reduction in nutritional value and solubility. Upon reconstitution with water some insoluble residue remains which would have to be removed prior to using the milk for infant feeding.

Spray draying represents an improvement over roller drying in that the dried product has higher nutritional value and has good solubility in water. Due to its bulkier nature the cost of packing, storing and transporting is slightly higher but an economic return is ensured on the considerable amounts of milk processed daily.

Often the milk is preheated to 63°C for 30 minutes or to just over 100°C for a few seconds. There is evidence that this produces anti-oxidants which combat the formation of rancid flavours in dried milk. The milk is then injected as a fine spray into the stainless steel spray drier shaped in the form of a tall inverted cone. Milk enters at the broad end where it meets a stream of hot air at 130–150°C. The droplets lose their water as they fall under gravity and collect as a fine powder in the narrow point of the drier. Because of the cooling reaction experienced by rapid evaporation the particles are virtually uncaramalised and the protein unaffected. In order to improve the solubility the powder may be slightly damped after it has passed through the spray drier and then redried, this restores physically bound water to the lactose sugar necessary for its solubility.

Eggs may be also successfully spray dried as an alternative to freezing. Care is needed in selecting liquid egg free from Salmonella contamination and pasteurisation, followed by early spray drying, is essential. Temperature control needs to be more precise than for milk if flavour is to be unimpaired. One of the chief disadvantages of egg powders is their instability which may result in decreased solubility. When this occurs certain functional properties of the egg are decreased. The successful preparation of cream puffs, meringues, sponge and angel cakes depends upon the ability of the egg protein to expand and hold air or steam. When deteriorated egg powders are used in place of whole eggs the product is inferior or even unacceptable. Badly deteriorated egg powder will not coagulate or thicken in batters, custards or dough products.

Tunnel drying incorporates passing the food on a moving belt which traverses a closed interior through which hot air can be blown by fans. Egg white can thus be produced as a thin film on a continuous belt system. Dried albumen can be used in white and chocolate cake mixes. Health foods based upon mixes of cereals and nuts can be commercially dried, economically, by control of the speed of traverse and the hot air temperature. Installation costs are less than with spray drying and the level of denaturation is less than with roller drying.

The type of drier to be used is largely determined by consideration of the chemical and physical nature of the product and the quantity to be processed. There is ample evidence to suggest that all three methods will continue to operate successfully with firms who have developed them to their maximum potential.

ANTIBIOTICS

Originally an antibiotic was regarded as a chemical substance produced by a micro-organism capable of destroying bacteria or preventing their growth. The most publicised was penicillin, the original forerunner of others such as aureomycin and streptomycin. Their use in foods is sometimes ironical. Penicillin is effective as an antibiotic for mastitis in cows but if it gets into the milk it can inhibit lactic acid bacteria so that cheese cannot be made. It is also possible for the milk to cause trouble in infants who are allergic to penicillin. However, these may be regarded as exceptional cases and the majority are considered to be harmless to humans in the amounts prescribed in food. At the correct level they impart no flavours or odours to food, and will not discolour the product.

Penicillin and streptomycin, when applied to the preservation of freshly caught fish, which are extremely perishable under storage aboard ship, produced no significant decrease in spoilage. The reasons lay in the narrow range of effectiveness against numerous spoilage bacteria and their relative instability in the food product.

Later research produced antibiotics capable of attacking spoilage micro-organisms on a broad front. Chlortetracycline at levels of 10–25 p.p.m. in ice extended considerably the preservation period for cod and halibut. Encouraging results were also achieved with meats and poultry. Beef carcasses treated with antibiotics can stand higher temperatures for the purpose of ageing and tenderising the beef. Microbial activity is delayed and this enables a wider time interval between slaughter and storage under refrigeration. Treatment of freshly dressed poultry with chlortetracycline extends its keeping quality and, in the amounts permitted for use in poultry, is destroyed during the cooking process.

Problems arise in the wholesale use of antibiotics in food. The protection they afford is limited, and within some period, ranging from a few days to a few weeks depending upon the temperature at which the anti-biotic-treated product is stored, the microbial flora will increase in the living animal. Most antibiotics have a limited bacteriostatic range; even the broad based ones are effective only against bacteria and are useless against yeasts and moulds. Indeed the growth of yeasts and moulds in antibiotic-treated beef appears to accelerate because of the lack of competition from bacteria. All antibiotics are relatively unstable; with few exceptions an effective concentration can be maintained in products for only a few weeks. This may however be an advantage in raw poultry because no residual antibiotic can be detected in the flesh after cooking.

ENZYMES

A major factor in food spoilage is the effect of enzymes upon consumable items. These may be naturally present in food causing autolysis or self-destruction.

In some foods partial autolysis is useful; for example, in meat tenderisation, but mostly they are undesirable particularly in high moisture content foods where they eventually cause total autolysis as in fish and fruit. Sugars will be completely converted to carbon dioxide and water. All enzymes may be destroyed by heat or inactivated by chemicals.

Maturation of the fruit or seed in vegetables or fruits (conversion of starch to sugar) is enzymic in nature. The ripening of bananas is one such application. Low temperatures only slow down enzymic activity and autolytic change must be halted in frozen foods by blanching prior to freezing. This involves heating to the destructive temperature for autolytic enzymes. Peas and beans will turn yellow if not blanched. Enzyme activity can take place slowly even at temperatures well below the freezing point of water because the enzymes, being adaptable proteins, can continue to work even when in a largely ice crystal matrix. This activity can be observed with temperatures as low as $-18°C$.

Some products have unusually high rates of autolysis of a proteolytic nature which is potentially hazardous to health because the compounds formed encourage the growth of pathogens. Examples include fish and shellfish where the pH is elevated by autolysis with liberation of tri-methylamine which encourages the growth of *Vibrio cholerae* and *Vibrio parahaemolyticus*.

Food discoloration is caused by autolysis, for example, browning of apples and potatoes when cut; soaking in brine inhibits discoloration. To prolong the storage life of fruit they may be placed in an enriched carbon dioxide atmosphere. However, the correct balance of oxygen/carbon dioxide must be maintained otherwise brown spot may develop. This is caused by anaerobic respiration which turns sugars into ethyl alcohol which then destroys tissue producing brown spots.

A frequent cause of rancidity in foods containing fat is due to lipase activity. Nuts, oils and milk require correct storage conditions if quality is to be preserved.

Enzymes may also be secreted by bacteria which have contaminated the food and eventually spoilage will result. Spoilage of a non-toxic kind involves a wide range of bacteria which multiply on food if conditions are suitable, producing total autolysis. Proteolytic degradation is the most probable leading to meat putrefaction. The slime formation commonly found on vegetables and in weak sugar solutions is of a dextrinous nature, initially due to bacterial activity. Micrococci in fats may produce rancidity. By contrast a useful bacterial spoilage produces cheeses of distinctive flavour. Lactic acid bacteria can act upon sugars to form lactic acid and diacetyl.

Since frozen foods continue to gain popularity, a blanching test has been devised to ensure good quality in peas, green beans, carrots, cauliflowers, green broccoli, sweet corn and potatoes. The test is the peroxidase enzyme test for enzyme survival in 'blanched' vegetables.

Peroxidase enzymes are very heat resistant. If they survive the heat treatment they can produce yellowing and 'off-flavours' in vegetables which have been frozen and stored at temperatures as low as $-18°C$. Activity is dormant at lower temperatures. To test for enzyme survival:

(a) Place a drop of 0.5 per cent guaiacol in 50 per cent ethanol and a drop of dilute hydrogen peroxide on the cut surface of the vegetable.

(b) Wait for about 5 minutes. A brown or orange colour indicates active peroxidase enzyme activity in the vegetable and that the 'blanching' heat treatment given has been insufficient to destroy the enzymes.

PREDICTIONS

It is to be expected that the convenience factor of food preserved by freezing will continue to increase its hold on a world market. Freezing does produce a slight deterioration in flavour and texture in food, and enzyme activity will still continue at a limited speed. Progress is being made on these lines to narrow the small gap in quality between frozen food and its freshly harvested equivalent. New techniques incorporating physical and chemical lines of approach may reduce deterioration during storage to a very low level so that food can be kept indefinitely for practical purposes.

The use of plastics for food packaging and wrapping is relatively new, and research will no doubt reveal novel variants on basic polymers such as polyethylene and polypropylene. Plastic films with sufficient heat resistance could replace tin cans as a container for sterilisation, offering a considerable saving in delivery costs. Ultrasonic vibrations of a frequency designed to destroy living bacterial cells in food would produce food in a safe condition without the need for heating. Radiation techniques are also capable of being extended to replace heat or refrigeration methods for food preservation.

Chemical additives represent another field for future research into improved food preservation. At present, large quantities of waste blood from abattoirs represent a big loss of valuable protein. By improving collection and preservation methods the marketing of blood for use in animal and human foods could become a large commercial activity.

Urea, ammonia and sodium metabisulphite have been tried as stabilisers; the latter at a 1 per cent concentration has produced stability of 28 days at 2°C, or 10 days at 30°C, without detectable microbiological or organoleptic changes occurring. The largest demand for blood products in the food industry comes from the manufacturers of prepared meats, who use blood plasma to increase the protein content and water-holding capacity of their products. However, the baking industry is interested in using plasma as a substitute for egg albumen, at a considerable reduction in costs.

Blood may be preserved by dehydration techniques involving the contact of blood with heated surfaces. This prolonged contact which is necessary to drive off most of the moisture tends to reduce the nutritive value of the final product. As a result the bulk of the dried blood produced in the United Kingdom is used as animal feed or fertiliser. It is possible that research will produce better stabilisers and these will in turn assist the development of better preservation techniques. Blood will then be subjected to new processes aimed at presenting it in a form acceptable as a meat analogue, in parallel with the soya bean which made its debut as 'Kesp'.

For the moment it seems unlikely that any major process is imminent to challenge the preservation offered by heat processing/aseptic packing and quick freezing. Efforts will be made to produce cheaper and possibly simpler methods of food preservation by using radiation, etc. This approach is dictated as a matter of survival, rather than to extend the range of sophisticated new products. The possession of atom bombs may have lessened the threat of a Third World War, but mankind has been presented with a novel problem—that of providing sufficient food of good quality for an ever increasing population. Food technology, it is to be hoped, will provide the answers, particularly in the field of preservation.

EXERCISE 18

1. What benefits are likely to arise by incorporating antibiotics as feed supplements for pigs? How can the results be explained?
2. Discuss the general principles upon which methods of food preservation are based.
3. What are the attractive features of food preservation through radiation sources?
4. What spoilage features are likely to be involved in refrigerated foods which would not be encountered in the canned equivalent?
5. Discuss possible hazards in the indiscriminate use of antibiotics as a means of food preservation.
6. In what ways can preservation alter the original qualities of the fresh product?

7. What factors do you consider are important in deciding the best way of preserving
   (*a*) meat and
   (*b*) fish?
8. Illustrate, by reference to appropriate sections of the food industry, the impact made by immersion refrigerants.

Suggestions for Further Reading

Abbott, A. F., *Ordinary Level Physics*, Heinemann Educational Books Ltd, London (1969).

Abbott, D., and Andrews, R. S., *An Introduction to Chromatography*, Longman Group Ltd, London (1970).

Callow, A. B., *Cooking and Nutritive Value*, O.U.P., London (1945).

Cameron, A. G., *Food Facts and Fallacies*, Faber and Faber Ltd, London (1960).

Carr, J. G., *Biological Principles in Fermentation*, Heinemann Educational Books Ltd, London (1968).

Collins, C. H., and Lyne, P. M., *Microbiological Methods*, Butterworth and Co., London (1970).

Drabble, J., *Textbook of Meat Inspection*, Angus and Robertson, London (1965).

Geary, D., 'Meat and Refrigeration', *Meat Trades Journal*, London (page 151, October 1968).

Gibbs, F. W., *Organic Chemistry Today*, Penguin Books Ltd, London (1970).

Griswold, R. M., *Experimental Study of Foods*, Constable and Co. Ltd, London (1969).

Gunstone, F. D., *An Introduction to the Chemistry of Fats and Fatty Acids*, Chapman and Hall, London (1968).

Heys, H. L., *Physical Chemistry*, G. Harrap and Co, London (1970).

Hildreth, E. M., *Elementary Science of Food*, Allman and Son, London (1969).

Mackinney, D., and Little, A., *Colour of Foods*, A.V.I. Publishing Co., London (1962).

Munro, J. W., *Pests of Stored Products*, Hutchinson Publishing Group Ltd, London (1966).

Muller, H. G., *An Introduction to Food Rheology*, W. Heinnemann Ltd, London (1973).

Pearson, D. E., *Laboratory Techniques in Food Analysis*, Butterworth and Co., London (1973).

Pyke, M., *Food Science and Technology*, John Murray Ltd, London (1970).

Rogers, J. L., and Binstead, R., *Quick Frozen Foods*, Food Trade Press Ltd, London (1972).

Schultz, H. W., *Food Enzymes*, A.V.I. Publishing Co., London (1960).

Sistrom, W. R., *Microbial Life*, R. and W. Holt, London (1969).

Answers to set problems

Exercise 2

1. 20.52
2. CH_4N_2O, $CO(NH_2)_2$
3. 607.1 kg
4. (*a*) 2.4 g   (*b*) 2.69 g
6. (*a*) 2×10^{-2} (*b*) 1 g ion/m^3 (*c*) 3
8. Heat of formation -12.4 kcal.

Exercise 4
7. Molecular weight 118 if the acid is assumed to be dibasic. Succinic acid $(CH_2COOH)_2$.

Exercise 5
7. 14 per cent. Acceptable quality, normally sausage 10 to 17 per cent protein content.
8. 4.4 per cent. Fondant would tend to be too brittle and of a coarse crystalline nature.

Exercise 7
1. 0.846 g mol
3. 4.3
5. 120.1
6. (*a*) 66.6 per cent   (*b*) 75 per cent.

Exercise 8
5. Monobasic, molecular weight 100.5
7. 91, oxalic acid $(COOH)_2$
8. 92, glycerol $C_3H_8O_3$

Past Examination Papers

INSTITUTE OF MEAT EXAMINATIONS

A complete set of science papers covering the various applications to the meat industry are illustrated to show the increase in standard from the first year (Craft Science) to the third year (Food Chemistry and Nutrition, Food Microbiology, Preservation and Hygiene). Craft Science has replaced the Elementary Science examination and the paper has altered considerably in both its style and number of questions to be answered. In the General Scientific Principles paper, which is attempted in the second year, a change has been made in the number of questions to be attempted. Prior to 1973 the choice was ten from twelve, now it is eight from eleven. The advanced papers remain as previous years, five questions to be selected from eight.

These changes have been designed to try to ensure a smooth transition from one year to another. A past criticism was that the first and second year papers were too close in standard, being below that expected from two years of studying meat science. By comparison the advanced year presented a much increased scientific knowledge of meat and meat products and a number of students were caught out by the increased tempo. Although there is room for improvement between the second and third year stages there is no doubt that the present arrangement represents a move in the right direction.

CRAFT SCIENCE 1973

Time allowed $1\frac{1}{2}$ hours. The paper is set in two Sections, A and B. Answer all *fifteen* questions in Section A and *five* questions only in Section B. You are advised to spend not more than forty minutes on Section A.

Section A. Answer all *fifteen* questions in the spaces provided or otherwise as indicated. Each question carries the number of marks indicated. Record your examination number in the spaces provided.

1. List three characteristics of living things that are common to both animals and plants (*a*) (*b*) (*c*). (3 marks)
2. (*a*) From where does the colour in animal fat originate? (*b*) What happens to this fat colour when animals are fed indoors during the winter months? (3 marks)
3. Name a nutrient found in meats which can be used for the following body functions: (*a*) body growth and repair; (*b*) provision of heat and energy; (*c*) manufacture of red blood cells. (3 marks)
4. (*a*) Name a water-soluble vitamin found in good supply in liver and kidneys.
   (*b*) Name a fat-soluble vitamin found in good supply in liver and kidneys. (3 marks)
5. What do you understand by the 'lacquering' of cans? (3 marks)
6. Draw a diagram of a second type (class) lever and clearly label the main parts. (3 marks)
7. (*a*) Name one insulating material used in the construction of a refrigerator.
   (*b*) Name one conducting material used in the construction of cooking vessels. (2 marks)

230

8. In each of the following cases state whether heat is taken in or given out
   (*a*) change of solid to liquid; (*b*) change of vapour to liquid; (*c*) change of liquid to vapour. (3 marks)
9. Convert 59° Fahrenheit to Centigrade. (3 marks)
10. Give two common sources of mould spores in a meat-processing area. (3 marks)
11. Complete the following sentence:
    Manufactured meat productions are easily contaminated with _____ and there-fore, during their production _____ conditions must be maintained. (2 marks)
12. State whether the following statements are 'true' or 'false':
    (*a*) Bacteria readily grow in dehydrated foods.
    (*b*) Refrigeration retards bacterial growth.
    (*c*) Pickled meats do not keep as well as fresh meats. (3 marks)
13. The recommended working temperature of a cold storage display cabinet for a range of fresh meat products is:
    (*a*) −18°C; (*b*) −5°C; (*c*) 2°C; (*d*) 10°C.
    Underline your choice of temperature. (2 marks)
14. Underline two foods in the following list which are often suspected when food poison-ing occurs:
    White bread; cold rolled joints of meat; chicken pies; smoked bacon; honey; pasteur-ised milk. (2 marks)
15. Name the acid present in meat which has formed from carbohydrate and aids in the slowing down of bacterial growth. (2 marks) Total marks—40.

Section B. Answer *five* questions *only* and in the answer book provided. Each question carries twelve marks.
1. (*a*) Define 'relative humidity'.
   (*b*) Describe how the relative humidity in a food preparation can be measured.
2. (*a*) Describe briefly what you understand by the process of digestion.
   (*b*) List the five main organs which are concerned with digestion.
3. (*a*) What chemical element is present in proteins that is not present in carbohydrates and fats?
   (*b*) Why is animal protein in the diet of greater value than vegetable protein?
4. (*a*) Give three common sources of bacteria in a vehicle used for transporting carcass meat.
   (*b*) Discuss what properties should be taken to keep contamination by these bacteria to a minimum.
5. (*a*) Name two salts necessary in a pickling 'brine' for the production of bacon.
   (*b*) Explain the principle on which a 'brineometer' works.
6. What type of packaging material would you use for the storage and display of (*a*) sliced bacon; (*b*) dripping? Discuss two properties for each material chosen and give the reasons for your choice.
7. (*a*) Why do hydrated food products have excellent keeping qualities?
   (*b*) Outline a method for the dehydration of a chosen food.
8. Give three personal hygiene rules that must be observed by meat handlers. For each rule give a reason why that rule is necessary.

General Scientific Principles 1973

Time allowed 2 hours. Answer *eight* questions only. All questions carry equal marks. Use diagrams and chemical formulae wherever possible.
1. Define the terms 'catalyst' and 'enzyme'. Briefly, how do enzymes tenderise meat? Name one enzyme found in the digestive tract; state the food component on which it acts and name the end product or products.
2. In the cleaning of utensils which have been used for the preparation of meat and meat products it is essential to employ soaps and synthetic detergents. Describe the main properties of these substances and explain detergent action.

3. A factor affecting the rate of drying meat is relative humidity. Distinguish between 'absolute relative humidity' and 'percentage relative humidity'. Draw and label an instrument which can be used for the measurement of relative humidity and explain the principles on which it works.

4. What is the essential difference between 'universal indicator paper' and 'narrow range indicator paper'?
   Describe in detail how you would use such papers to find the pH of meat.

5. Distinguish between a.c. and d.c. and give an example of each. List five safety rules which should be observed when using electricity.

6. Briefly discuss the chemical make up of meat proteins and describe their important physical and chemical properties.

7. Meat is an expensive commodity. On what nutritional grounds is meat still a good buy economically?

8. Distinguish between temporary hard water and permanent hard water. Give one method of softening temporary hard water and include a chemical explanation. State one disadvantage of hard water to the meat industry.

9. Write short notes on the chemical and physical properties of any one of the constituents of air, emphasising any effects that it may have on meat.

10. Contact by certain metals may be harmful to meat. Name three such metals and for each indicate its effect on the product.

11. Draw and label a diagram of the Carbon Cycle. Explain the cycle and show how meat animals form a part of it.

Food Chemistry and Nutrition

Time allowed 3 hours. Five questions to be answered. All questions carry equal marks. Use diagrams and graphs wherever possible.

1. The mineral content or 'ash' of lean meat is approximately 1 per cent of the total weight. What are the main mineral constituents of this ash and from which tissues of the animal do they generally derive? What information can be deduced from estimation and analysis of the ash content of a manufactured meat product such as sausage?

2. What is the relationship between Vitamin A and beta-Carotene?
   Discuss the factors affecting the levels of these two substances in meat and indicate the general role of Vitamin A in the nutrition of man.

3. Summarise, in your own words, the requirements of the Sausage and Other Meat Product Regulations (No. 862) 1967 in respect of the minimum total meat content of sausages.
   Give brief details of the procedure for determining the total meat content of sausages.

4. What useful information can be gained from a record of the post-slaughter pH changes occurring in a carcass?
   Since polyphosphates tend to raise and stabilise the pH of meat what benefits are obtained from their use in the manufacture of products such as sausages and ham?

5. Discuss the various factors that affect the natural flavour of meat.

6. What is meant by 'oxidative rancidity'?
   Describe the conditions which promote this type of rancidity and give brief details of any laboratory test which indicates either:
   (a) the susceptibility of fat to turn rancid; or
   (b) the actual degree of rancidity attained.

7. Write short notes on the following:
   (a) Calorific value; (b) lecithin; (c) first class protein (high biological value protein);
       (d) deficiency diseases.

8. Discuss the ways in which the establishment of a quality control laboratory with a large meat products manufacturing company could aid the achievement of:
   (a) greater economy of production;
   (b) improved marketing and sales.

Food Microbiology, Preservation and Hygiene

Time allowed 3 hours. Five questions to be answered. All questions carry equal marks. Use diagrams and graphs wherever possible.

1. Describe the various phases that occur during the growth of bacteria under optimum conditions.
   What environmental factors influence this pattern to cause sub-optimal growth and how far are these factors applicable to food preservation?

2. Regulation 27 of the current Food Hygiene Regulations demands a measure of temperature control for perishable foodstuffs (including certain meats) whereby foods intended for catering services must be kept either below 50°F (10°C) or above 145°F (63°C). It has been suggested that this Regulation be extended to cover similar foodstuffs on retail display.
   Discuss the effect of such an extension in the control of food poisoning and food spoilage of meat products. Are there any practical disadvantages to this proposal?

3. Give accounts of two methods employed for the counting of bacteria.

4. Under what circumstances may mould spoilage of foodstuffs displace the more usual, bacterial-type spoilage?

5. Manufactured meat products have been associated with outbreaks of both food-borne and food poisoning diseases. What are the various differences between these two forms of disease, including (briefly) control measures?

6. In what major respects does the canning of pasteurised ham differ from ordinary canning procedures?
   What special, post-processing precautions are required for the safe handling and display of pasteurised canned products and why?

7. Discuss the advantages and disadvantages of assessing the safety of a food product in terms of:
   (*a*) the total numbers of bacteria present;
   (*b*) the various types of bacteria present.

8. Describe two forms of food preservation processing that do not utilise any form of temperature regulation or heat application.

ORDINARY NATIONAL DIPLOMA IN FOOD TECHNOLOGY (second year)

The science content for the second year stage presupposes *either* a satisfactory completion of the first year stage *or* entry to this stage of the course by virtue of other qualifications deemed by the Committee to be equivalent, for example industrial experience and academic expertise.

Throughout the whole two-year course practical work in all appropriate subjects will be examined by a system of continuous assessment.

There are four science papers set. These are marked internally and then assessed externally by a moderator of high academic and practical ability in the paper or papers for which he is responsible. The four papers are as follows:

Chemical and Biological Sciences
Paper 1. Chemistry, Biochemistry and Nutrition.
Paper 2. Biology, Microbiology and Hygiene Food Science.

Technology
Paper 1. Scientific Principles of Food Processing.
Paper 2. Quality Control and Food Properties.

It must be emphasised that correct use of the library and current periodic journals dealing with food in general should be consulted if the questions set are to receive satisfactory answers. The four papers set for 1973 are illustrated below.

Chemistry, Biochemistry and Nutrition

Time 3 hours. All questions carry equal marks. Answer *five* questions, at least *two* from each section.

Section A

1. To what extent may enzymes be regarded as catalysts? Your answer should be illustrated by reference to the manufacture of bread and metabolism of food nutrients.
2. What advantages are offered by artificial, over natural, flavours? How can flavour be controlled on a pilot scale to ensure, as far as possible, consistency in bulk production?
3. Discuss the chemical basis of colour. Why is the colour of fresh meat more prone to change than cured, and indicate briefly the colour requirements for the soft drink, sweets and canning industries.
4. What do you understand by the terms:
   (*a*) oxidising agent; (*b*) hydrolysing agent; and (*c*) proteolysis?
   How can the quality of bakery and butchery products be related to a practical knowledge of all three items?
5. 'The Law of Mass Action is fundamental to the food industry'. What evidence is there for this statement?

Section B

6. Distinguish between oxidative rancidity and hydrolytic rancidity in oils and fats. Include in your answer equations and mechanisms wherever possible. Indicate measures which could be taken to reduce: (*a*) oxidative reducity; and (*b*) hydrolytic rancidity in food.
7. You are presented with two white powders—one of which is glucose, the other lactose. Give the principles, and equations where possible, relevant to two methods of distinguishing between the two sugars. Organoleptic tests should not be included. Draw the pyranose structure of alpha D glucose.
8. An elderly couple are reported to be suffering from malnutrition. Assuming that you have unrestricted access to them, how would you investigate their nutritional levels? Considering present-day circumstances, what defects might you find in an old-age-pensioner's diet—explain your answer.
9. Write short notes on:
   (*a*) ATP and the metabolism of glucose; and
   (*b*) Fibre and the diet.

Biology, Microbiology and Hygiene

Time 3 hours. This paper is divided into *two* sections. Answer *five* questions, at least *two* questions from each section. Wherever possible answers should be illustrated by suitable examples and diagrams. All questions carry equal marks.

Section 1 Food Microbiology
Question 1
An outbreak of skin infection on the hands of slaughter-house workers has occurred at a meat products factory. Describe the investigation you would carry out, with particular attention to the possible dangers of enterotoxin development in products.

Question 2
A routine surface evaluation indicates high mould spore levels in a plant bakery. What measures would you recommend to eradicate these micro-organisms and discuss the possible dangers of mycotoxins.

Question 3
What are the analytical methods used to positively identify faecal streptococci? Discuss the significance of these bacteria in meat products.

Question 4
Distinguish between Fermentation and Oxidative Yeasts giving examples of types found in food products. Show how spoilage types can be controlled.

Question 5
Write brief notes on the following:
(*a*) *Clostridium welchii* (perfringens);
(*b*) Mannitol salt agar;
(*c*) *Aspergillus flavus*; and
(*d*) *Rhodotorula gracilis*.

Question 6
Write a technical essay on: 'The microbiology of liquid milk'.

Section 2 Food Hygiene and Legislation
Question 7
Describe the methods used to control insect pests in food factories. Comment on their relative merits.

Question 8
Write brief notes on the following:
(*a*) hypochlorites;
(*b*) air curtains;
(*c*) insect traps; and
(*d*) electrostatic precipitators.

Question 9
Describe the cleansing routines and chemical agents used in a vegetable freeze-processing plant.

Question 10
Unsound food may be passed on to the public for human consumption. Briefly outline the proceedings used to prevent this from happening.

Question 11
Why is food legislation necessary? How is it enforced?

Question 12
The Milk (Special Designation) Regulations 1963 and Amendment Regulations 1965 are partly concerned with the heat treatment of milk. What is the objective of this heat treatment? List the types of process concerned with the corresponding standards of processing.

Question 13
Some foods intended for human consumption should be kept within controlled ranges of temperature. What piece of legislation requires this? What are the temperatures and what are the foodstuffs concerned? Why is it desirable with modern food production techniques to have such requirements?

Scientific Principles of Food Processing

Time 3 hours. All questions carry equal marks. Answer *six* questions, at least *two* from each section. Wherever possible answers should be illustrated by suitable examples and diagrams.

Section A

1. Discuss the production problems of canned fruit and canned meat, and outline the scientific principles governing their production.
2. Write a technical essay on 'The packaging of frozen foods'.
3. How can the food technologist make use of dehydration? Discuss the advantages and disadvantages of this method of food preservation.
4. Discuss the special production problems which are encountered with the following food products, and show how these are solved:
   (*a*) meat pies;
   (*b*) frozen desserts;
   (*c*) canned vegetables; and
   (*d*) dried soup-mixes.
5. Given an account of the methods of gas packing, illustrating your answer with examples of food industry applications.
6. The use of Ionising Radiations for sterilising foods presents many problems for the Food Technologist. Outline the major problems and suggest how these may be solved.

Section B

7. Explain fully what you understand by *four* of the following:
   (*a*) elutriator;
   (*b*) precoating;
   (*c*) reflux ratio;
   (*d*) seed crystal;
   (*e*) saturated steam; and
   (*f*) thickener.
8. How is Reynolds Number *Re* calculated in given circumstances?
   What is the significance of the *Re* value in:
   (*a*) the layout of process piping and the choice of pipework components;
   (*b*) the design and operation of any kind of fluid heat exchanger?
9. Describe the action of tangential flow, axial flow, and radial flow agitators and discuss their relative merits in liquid mixing. Describe the type of mixer (and agitator) you would choose to affect the mixing of:
   (*a*) a viscous dough or paste;
   (*b*) a dry powder blend; and
   (*c*) an oil/water mixture prior to transfer to an emulsifier.
10. With the aid of sketches, describe the construction and mode of operation of a wiped film evaporator. What advantages do film-type evaporators offer, especially in the food processing industry?
11. A swinging hammer is to be used to reduce grain rice to a powder. Sketch this type of mill and explain why it may usually be said to operate under 'closed circuit' conditions. A sieve analysis of 100 g of product gave these results:

	I	II
Retained on 100 mesh	nil	0.5 g
Retained on 150 mesh	80.1 g	80.1 g
Retained on 200 mesh	12.4 g	12.2 g
Retained on 300 mesh	5.1 g	4.9 g
Passing 300 mesh	2.4 g	2.3 g

Using the figures of Column I calculate the production rate of specification ground rice 'passing 100 mesh, retained on 100 mesh', if the rice grain feed rate is 250 kgf/hour. If, later in the week, sieve analysis II is observed, what would you suspect has happened?

12. What factors affect the drying rate of solids (with special reference to food materials)? Describe the construction and mode of operation of *either* a drier *or* a fluidised bed drier. State what kind of material would be most suitable for the drier of your choice.

Quality Control and Food Properties

Time 3 hours. Answer *six* questions, at least *two* from *each* section. All questions carry equal marks.

Section A

1. The analysis of pork sausages gave the following results:
Fat 37.5 per cent, Water 42.0 per cent, Ash 1.6 per cent, Nitrogen 1.282 per cent. The sausages contained no soya, no S.M.P. and microscopical examination revealed muscle fibres and wheat starch. Calculate the meat content, perform a check calculation and then comment on the meat content in relation to The Sausage and Other Meat Products Regulations.
2. Included in the Milk (Special Designation) Regulations are the Methylene Blue Test and the Turbidity Test. Give the principles behind each test and state what information can be drawn from the results. Draw and label a lactometer.
3. In quality control, what is meant by 'Sampling'? Describe in detail the method of sampling that you would employ for: (*a*) Sausage; (*b*) Shredded Suet; and (*c*) Butter.
4. Give the legal requirements concerning the quantities of chalk to be added to plain white flour. State clearly the principles behind each part of the experiment that you would perform to check the legislation. Show clearly the calculations that would be required.

Section B

5. Discuss the problems of Quality Control/Assurance in the use of fruit and vegetables raw materials.
6. Describe the range of dairy products available for the industrial user giving examples of applications.
7. Discuss the types of starches used in food production stating the applications of each type.
8. What factors influence the choice of meat cuts for: (*a*) soup production; (*b*) steakburger production; and (*c*) beef pie production.

National Diploma in Baking (second year)

The paper is broadly based in its approach to scientific principles and may include questions relevant to the first year. Answers should include, wherever possible, any industrial applications.

A practical examination must also be passed in applied food science. Attention is drawn to the candidates for marks awarded on a good presentation basis.

Applied Science (Theory)

Time 3 hours. Answer any *six* questions. Sixteen marks are awarded for each question and four marks for overall presentation of the paper. Formulae, equations and diagrams should be used wherever possible.

Question 1

Give three properties of acids. Name any four acids, other than acetic acid, encountered in the baking industry and say in what connection each is found.

Calculate the percentage of acetic acid (CH_3COOH) in vinegar if 2 ml of vinegar require 14 ml $\frac{N}{10}$ NaOH for neutralisation (atomic weights: Na...23, C...12, O...16, H...1). Describe briefly how you would find the approximate pH of a solution.

Question 2

Differentiate between a simple triglyceride and a mixed triglyceride. Why does lard not have a sharp melting point? Describe briefly how you would find the melting point of a fat.

Question 3
Differentiate between oxidation and oxygenation using suitable examples. Give the chemical equation for the burning of propane (C_3H_8) in an unrestricted supply of oxygen. An oven on a 250 volt main is rated at 20 amps. How much will it cost to operate the oven for 10 hours if electricity costs 0.8 p per unit? Define: (*a*) A British Thermal Unit; and (*b*) Therm.

Question 4
Give five factors which affect the rate of drying of a food. Describe briefly a method for determining the percentage of moisture in flour. What result would you expect from normal white flour?

Question 5
What quantity of chalk has to be added to plain white flour under present legislation? Give the scientific principles behind an experiment designed to check this legislation.

Question 6
What is the relationship between the kilocalorie and the kilojoule? Describe fully the three main uses of energy by the body. Suggest two foods which have a high energy content and say why they have this property.

Question 7
(*a*) Define Relative Humidity and Percentage Relative Humidity.
(*b*) Give the chemical formulae of two amino acids.
(*c*) Define Iodine Value. What does it measure?
(*d*) Name a hydrometer used in the bakery. What does it measure?
(*e*) Name a mineral found in food. What is its function in the diet?
(*f*) Name two monosaccharides.
(*g*) Name two polysaccharides.
(*h*) Name two enzymes found in the digestive tract.

Question 8
Compare and contrast the conditions necessary for the growth of bacteria and moulds. How can bakery foods be processed and stored to prevent growth and development of such micro-organisms?

Question 9
'Hygiene Education is one of the most important factors in ensuring a safe bakery trade'. Discuss the relevance of this statement and the main points that would be covered in such Hygiene Education.

Index